MILITARY REFORM AND DEMOCRATISATION
Turkish and Indonesian experiences at
the turn of the millennium

KARABEKIR AKKOYUNLU

ADELPHI PAPER 392

The International Institute for Strategic Studies

Arundel House | 13–15 Arundel Street | Temple Place | London | WC2R 3DX | UK

ADELPHI PAPER 392

First published November 2007 by **Routledge**
4 Park Square, Milton Park, Abingdon, Oxon, OX14 4RN

for **The International Institute for Strategic Studies**
Arundel House, 13–15 Arundel Street, Temple Place, London, WC2R 3DX, UK
www.iiss.org

Simultaneously published in the USA and Canada by **Routledge**
270 Madison Ave., New York, NY 10016

Routledge is an imprint of Taylor & Francis, an Informa Business

© 2007 The International Institute for Strategic Studies

DIRECTOR-GENERAL AND CHIEF EXECUTIVE John Chipman
EDITOR Patrick Cronin
MANAGER FOR EDITORIAL SERVICES Ayse Abdullah
ASSISTANT EDITOR Katharine Fletcher
PRODUCTION John Buck
COVER IMAGE Getty

Printed and bound in Great Britain by Bell & Bain Ltd, Thornliebank, Glasgow

British Library Cataloguing in Publication Data
A catalogue record for this book is available from the British Library

Library of Congress Cataloging in Publication Data

ISBN 978-0-415-46443-7
ISSN 0567-932X

Contents

GLOSSARY

AKP	Justice and Development Party (Turkey)
ASEAN	Association of Southeast Asian Nations
DPD	Regional Representative Council (Indonesia)
DPR	People's Representative Council (Indonesia)
MPR	People's Consultative Assembly (Indonesia)
PKK	Kurdistan Workers' Party
SSR	security-sector reform
TNI	Indonesian Armed Forces
TSK	Turkish Armed Forces

INTRODUCTION

'To a man with a hammer, everything looks like a nail.'

– Mark Twain

Military reform is an integral part of a country's democratisation. The 'normal' role of the military in a democratic country, as the armed branch of the state apparatus, is to defend the country against external armed aggression under the authority of democratically elected civilian policymakers. A country cannot be considered fully democratic unless the military in particular, and society in general, accept and internalise this normative role.

Working from this fundamental premise, this paper looks at military reform and democratisation through the experiences of Turkey and Indonesia at the turn of the millennium. More specifically, it endeavours to understand the various factors – both internal and external – that affect the nature and the outcome of military reform initiatives within the process of democratisation, based on its analysis of the two countries.

Although it aims to make general observations based on the two case studies, the paper acknowledges that there is no fixed recipe for successful military reform and democratisation that can be applied to all countries. Instead, every country presents unique factors that influence the fate of its democratic reforms, and therefore has to be evaluated first and foremost within its specific context. To this end, this paper analyses each case study separately within three main categories – society, the global order and the state – and looks for answers to questions such as:

- What is the traditional role of the military in society; and what social, cultural and historical characteristics explain this role?
- How is the military perceived in society, and what factors shape this perception?
- What are the societal and military attitudes towards the concepts of military reform and democratisation?
- What are the regional and global trends, events or actors that promote or obstruct democratisation in each country?
- What dynamics among the different branches of the state apparatus – such as the military, the executive branch, the legislature, etc. – explain the changes in civil–military relations, and the success or failure of democratic reform initiatives?

It should be noted that, when dealing with roles, perceptions and attitudes, this paper focuses on their evolving nature, and does not treat them as static factors. Only after thoroughly analysing each country in its unique and dynamic context can we compare and contrast their experiences, and make conceptual observations about military reform and democratisation, and the factors that influence them.

The paper is divided into four chapters. The first chapter sets the theoretical framework of the debate on military reform and democratisation and discusses how theory will be applied to the analysis of specific cases. The second and third chapters focus on Turkey and Indonesia respectively. They look at the reform initiatives in both countries in light of the factors that influence them at the levels of society, the global order and the state. The fourth chapter engages in a comparative analysis of the two case studies, and goes on to conclude with some general observations that can assist the study of military reform and democratisation.

Why Turkey and Indonesia?

There are considerable similarities between the factors that have shaped the almost simultaneous attempts at military reform and democratisation in Turkey and Indonesia. Both Turkey and Indonesia have militaries that play a traditionally central socio-political role, and which have been the object of reform initiatives within a wider context of democratisation at the turn of the millennium.

This resemblance becomes more significant when one considers that both Turkey and Indonesia are non-Arab countries with predominantly Muslim populations. At the same time, both societies are often identified as 'moderate' in religious fervour, hosting a secular regime, rather than one based on Islamic laws and teachings. Both are considered to be developing countries, which, despite experiencing substantial growth in their economies over the past two decades, still exhibit wide income discrepancies, with poverty continuing to plague a significant proportion of both societies.

The Turkish and Indonesian militaries have been continually involved – in a direct or indirect fashion – in the political affairs of their countries since the inception of both republics. Historically situated at the core of the regime's power structure, both militaries have an acute and rather simplistic threat perception, categorising any perceived challenge to the status quo as a threat to the regime. While the most serious traditional threats for both militaries remain political Islam and ethnic separatism, with the end of the Cold War, the economically and politically liberalising forces of globalisation have also been identified as potential threats to the unity of the nation-state and the preservation of the status quo.

It is not only their similarities that make a comparison of these two countries interesting; it is also their differences. Bridging Europe to the Middle East and the rest of Asia, Turkey finds itself in a strategically important, and particularly sensitive, geopolitical position. A key NATO ally – providing the second largest military force to the Alliance – and a candidate for European Union membership, Turkey has been a multi-party democracy for over 50 years, despite brief but regular interventions by its military. Turkey's most comprehensive push for democratisation and military reform to date began at the end of the 1990s, as a result of a mounting popular drive for EU membership, and reached its peak under the government of the 'mildly' Islamist Justice and Development Party (AKP), which, despite breaking from its fervently Islamic political heritage and moving to the liberal mainstream, is still regarded as a sinister threat to Turkey's secular regime by the military and the bureaucratic elite.

In the case of Indonesia, the considerable regional tranquillity that this founding member of the Association of Southeast Asian Nations (ASEAN) has enjoyed due to its relative geopolitical isolation is often overshadowed by the immense task of keeping a sprawling archipelago and diverse population intact under a single nationhood and central government. The former Dutch colony, united for the first time in history as a single and sovereign entity after the Second World War, was ruled by two authoritarian leaders – Sukarno followed by Suharto – between 1949 and 1998. In 1965, an anti-communist coup against Sukarno's regime, led by General Suharto, established a military-backed dictatorship that lasted for over three decades. The military regime, known as the New Order, collapsed in the face of a major economic crisis and popular riots in 1998, triggering a chaotic yet energetic era of democratic transition and military reform. The first free and fair election in Indonesia's history was held as recently as 1999.

Most studies of democratisation limit their analyses to particular regions, such as Eastern Europe or Latin America. Without questioning the practical wisdom of the regional approach, this paper offers an alternative way of looking at military reform and democratisation. In bringing together findings from two countries of different regions it aims both to demonstrate vital regional differences and to emphasise the aspects of military reform and democratisation that transcend regional boundaries.

CHAPTER ONE

Setting the Theoretical Framework

Military reform and democratisation as intertwined processes

This paper assumes that military reform and democratisation are co-directional processes: they share certain goals, and one process does not counter the other. They are, furthermore, highly interconnected, and it is difficult to talk about one outside the context of the other. Processes of democratisation very often involve substantial reforms of a country's military establishment. Likewise, military reforms are often geared towards achieving greater accountability and transparency within the security sector, which in turn reflects a wider push for democracy in a society. This paper may thus be read both as an examination of military reforms with a focus on democratisation, and as a study of democratisation through the area of military reform.

In an environment of changing security needs and socio-political demands, democracies and militaries come under constant pressure to evolve and adapt. Military reform and democratisation should therefore also be regarded as continuous processes, rather than as end points in themselves.[1]

Democratisation

Democratisation is generally understood as the process undertaken to achieve democracy, or more specifically *liberal* democracy – as opposed to socialist, participatory or community-focused democracy.[2] During the Cold War, in the West, 'democracy' became synonymous with 'the free world', which described the political arrangements in non-communist Western countries. Embracing liberal economic and capitalist ideas, 'democracy came to mean almost exclusively liberal or representative democracy … the empirical "reality" of the West'.[3]

While acknowledging its practical shortcomings and ideological affiliations, this paper refers to the liberal representative kind of democracy as the object of democratisation, as this is still the most widely accepted and internationally promoted variant, long after the end of the Cold War. In the words of political scientist Laurence Whitehead:

> The point here is not to debate the relative merits of rival viewpoints, but to emphasize that, when the leading liberal capitalist governments attempt to promote democracy internationally, it is the first [liberal representative] variant, that meets their strongest approval, and that is seen to pose the least threat to their real interests.[4]

We should, however, distinguish between liberal democracy and a 'liberalising' authoritarian regime. Whitehead and Schmitter note that powerful minorities in authoritarian regimes often find it advantageous to adopt 'at least an appearance of sympathy for democracy' in order to defend and preserve their privileges.[5] This is pertinent today to liberalising non-democracies such as China.

While democratisation typically involves various liberal reforms aimed at the state and society, there is no guarantee that liberalisation will lead a non-democratic country to full democracy. According to political science scholar Alfred Stepan:

> In a non-democratic setting, *liberalization* may entail a mix of policy and social changes, such as less censorship of the media, … perhaps measures for improving the distribution of income, and most important, the toleration of opposition … *Democratisation* entails liberalization but is a wider and more specifically political concept. Democratisation requires open contestation over the right to win control of the government, and this in turn requires free competitive elections, the results of which determine who governs.[6]

Here, Stepan identifies the primary practical goal of democratisation: 'open contestation over the right to win control of the government', or 'free competitive elections'. Samuel Huntington refers to free and fair elections as 'the essence of democracy, the inescapable sine qua non'. He argues that freely elected governments may be 'inefficient, corrupt, short sighted, irresponsible', but that these qualities do not make such governments undemocratic.[7]

Some scholars object to such minimal criteria and look for broader bases for defining the goals of democratisation. According to political scientist David Held, a 'proper democracy' must acknowledge the importance of an 'impersonal structure of public power, of a constitution to help protect and safeguard rights, and of a diversity of power centres within and outside the state … to promote open discussion and deliberation among alternative political viewpoints and platforms'.[8]

Mary Kaldor and Ivan Vejvoda distinguish between 'formal' and 'substantive' democracies. They define formal democracy as 'a set of rules, procedures and institutions', whereas substantive democracy is 'a process that has to be continually reproduced'.[9] On a continuum, international politics scholar Jean Grugel places formal democracy at the minimalist end, and substantive democracy at the maximalist end. 'The litmus-test for democracy', she writes, 'is not whether rights exist on paper but, rather, whether they have real meaning for people. Inevitably, this implies a redistribution of power.'[10]

This paper holds that democratisation is an ongoing process, with the ultimate goal of establishing substantive, or proper – rather than formal, or minimal – democracy. This is the measure that will be employed in analysing the process and the outcome of Turkish and Indonesian democratisation efforts.

Military reform

The exclusive duty of the military in a 'normal' democracy is to defend the country against foreign aggression under the political authority of the elected government. In some countries, however, powerful militaries function beyond this role, and engage in policymaking and law enforcement. The fundamental objective of military reform, in the context of democratisation, is to reorient and limit the military's activities to its normative duties, and subordinate it to those democratically elected to take charge of the country's affairs.[11]

Civil–military relations were a subject of particular scholarly interest during the Cold War. The theories of two analysts above all, Samuel Huntington and Morris Janowitz, set the parameters of the debate for several decades. Huntington, in his classic *Soldier and the State*, argued that in order to reduce military involvement in politics, military professionalism (i.e., officers' education, sophistication and specialisation) had to be maximised. 'A highly professional officer corps', he wrote, 'stands ready to carry out the wishes of any civilian group which secures legitimate authority within the state'.[12] Janowitz, in *Professional Soldier*,[13] countered that professionalism rather enhanced the chances of military involvement in civilian affairs, 'especially in those cases where the civilian institutions were found to be … underdeveloped and civil culture lacking'.[14]

The end of the Cold War brought new questions to the fields of security studies. It became obvious that security could no longer be sufficiently understood in military terms alone. New issues, such as environmental problems, disease, drug and arms trafficking and terrorism, whose main actors are often neither the military nor the state, came to be perceived as security threats. Added to this was the spectacular challenge of achieving and consolidating democratic reform in the ex-communist countries, where the security sector was seen as 'one of the residual elements of the old regime'.[15]

The concept of 'security-sector reform' (SSR) developed as a response to these challenges, combining as it did the themes of civil–military relations and defence reform. Although it is still a nascent and ill-defined subject – hardly picked up by American scholars – the thinking on SSR has usefully challenged conventional ways of studying the security sector and civil–military relations.

Political scientist Timothy Edmunds regards SSR as a 'holistic approach', which recognises that 'the role of security and security sector actors in political and economic reform is important and complex, and not simply limited to questions of military praetorianism and civilian control over the armed forces'. The goal of SSR should be 'the provision of security in a state in an effective and efficient manner, and in the framework of democratic civilian control'.[16]

This paper draws substantially on the SSR literature, especially on country-specific works. However, it is not a study of SSR, and the scope of the paper is considerably different. It is primarily concerned with the military, and therefore will not focus on judicial reforms or private security services, areas that have been examined in some detail by scholars working to a broad definition of SSR. It is also chiefly concerned with military reform in the context of democratisation, and therefore economic development, also an area of interest to many SSR specialists, will be considered only when directly related to this concern.

Some readers may infer from the assumption that military reform and democratisation are co-directional that the military must by its very nature be against reform or democratisation. Is the military a natural obstacle on a country's path to democracy? This paper holds that it is not. Reform almost always meets opposition, and established and powerful institutions can be notoriously reluctant to change.[17] However, military reform is not *against* the military, just as the military does not have to be against democracy. Some of the leading democracies in the world also command exceptional militaries.

In some cases, the military has even been the instigator of democratic reforms, for instance in Brazil and Portugal.[18] Linz and Stepan suggest that 'all hierarchical military regimes share one characteristic that is potentially favourable to democratic transition':

> The officer corps, taken as a whole, sees itself as a permanent part of the state apparatus, with enduring interests and permanent functions that transcend the interests of the government of the day. This means that there is always the possibility that the hierarchical leaders of the military-as-institution will come to the decision that the costs of direct involvement in non-democratic rule are greater than the costs of extrication … Paradoxically but predictably, democratic elections are thus often a part of the extrication strategy of military institutions that feel threatened by their prominent role in non-democratic states. [19]

In both Turkey and Indonesia, our two case studies, the military has, at certain points in recent history, acted in ways that would support the above observation.

Successful democratisation often nevertheless entails a thorough restructuring of the security sector, as a substantive (as distinct from a formal) democracy calls for a military that is accountable to the democratically elected civilian government. This is the main objective of military reform. To sum up in the words of politics and security analyst Theodor Winkler:

> Democracy cannot develop if the population knows that the security apparatus is not under firm political and democratic control, but able to strike out independently at any given moment – be it by putting pressure on the government, by exempting its own budget, structures, procurement processes, infrastructures and personnel from democratic scrutiny, or even by toppling the government.[20]

Democratisation in cases where the military has such powers will usually necessitate a change in the way the military's role and position are perceived within the state and society.

A three-layer approach to the study of military reform and democratisation

Although there is no universal recipe for successful democratisation, it can be useful to look at democratisation as it occurs in different contexts in terms of common categories, with a view to drawing some tentative comparative conclusions.[21] Several factors play an important role in determining the success of military reform in any democratising country, such as dynamic internal and external political, economic and security conditions, timing (domestic and international trends), culture, history and the willingness of the reformers. Borrowing from Grugel, this paper will group these dimensions of change into three core categories, or layers of influence, which will be used in looking at our case studies: society, the global order and the state.[22]

Society

'Society' is the inside layer, and refers to domestic factors such as sociopolitical trends, culture and history. How the military perceives itself within society is examined in this section, as well as how society perceives the position and the role of the military, and how both military and society perceive the concepts of military reform and democratisation. These perceptions may have developed as a result of cultural characteristics or history, or in reaction to trends and developments on the state and global levels. Social perceptions of the military can be crucial determinants of the success of democratisation. As the case studies will demonstrate, in some countries the military's sacrosanct self-perception may be largely approved and upheld by society, which can make the military more hostile towards foreign and domestic criticism and reluctant to reform.

In most democratising non-Western countries, both democratisation and military reform are often portrayed as 'foreign' ideas. Societies react differently to outside influences. Global ideas and trends may trickle down through a society – often via the elite – but how that society perceives and reacts to them depends on particular cultural and historical characteristics.

Furthermore, the perception of and reactions to outside ideas of the military elite, which may or may not reflect those of the society at large, also influence the readiness with which these ideas are adopted. Political analyst Emily Goldman argues that when elites are committed to 'defending an official orthodoxy based on an ideological belief in the corrosive influence of foreign ideas, they respond to crises in ways that preserve orthodoxy and this limits diffusion'.[23]

In short, the military cannot be studied separately from society. Likewise, 'what is happening in the military will tell a lot about what is happening in the society'.[24]

The global order

The 'global order' represents the outermost layer in this analysis. It covers the contribution of global and regional[25] actors, and the evolving environment of international politics, economics and security, to determining the energy with which democratisation and military reform are undertaken by governments, and supported or resisted by societies and militaries. States and societies, being parts of the global order, are directly influenced by it, and in turn they influence the global order. This interconnectedness has been greatly enhanced, especially since the end of the Cold War and the triumph of liberal economic values, by the increasing integration of markets and advancements in telecommunications, i.e., globalisation.

Numerous arguments attempt to explain when and why countries democratise. A popular theory was put forth by Samuel Huntington in the early 1990s. In *The Third Wave*, Huntington suggests that democratisation happens in 'waves', that is, as a global trend whose popularity surges from time to time. According to Huntington, there were three such waves of democratisation between the early nineteenth century and the early 1990s. Each of the first two waves was followed by what he calls a 'reverse wave', when a number of democracies reverted to authoritarian rule.[26]

The latest wave – which kicked off with the democratisation of Portugal in 1974, continued with Spain, Greece and parts of Asia and Africa, and peaked with the disintegration of the USSR in 1991 – had several root causes, including the deepening legitimacy problem of authoritarian systems, unprecedented global economic growth beginning in the 1960s, advances in the telecommunications industry and the 'snowball' effect of demonstrations, as well as an increased emphasis on democratisation and human rights on the part of international actors such as the European Council and NATO.[27]

Subscribing to Huntington's wave theory, one can argue that this third wave flowed straight through the 1990s and into the new millennium. The end of the Cold War, the collapse of communism in Europe, the democratisation of Central and Eastern European republics and finally the enlargement of the European Union to 27 countries may be considered a horizontal continuation of the third wave, and a triumph for Western liberal democracy. Indeed, Francis Fukuyama declared 'the end of history' as a result of the emergence of the democratic United States as the only superpower in the world.[28]

Yet the same period also saw a sharp increase in civil conflicts and terrorism. Although it is still perhaps too early to judge, the events following the 11 September 2001 attacks suggest that 'the triumph of democracy' has been undermined. Huntington wrote that 'the United States is the premier democratic country of the modern world, and its identity as a nation is inseparable from its commitment to liberal and democratic values'.[29] Following the US debacle in Iraq, and the prospect of possible similar failure in Afghanistan, as well as the well-publicised human-rights abuses in Abu Ghraib and Guantanamo prisons, it would be harder to make this statement with confidence. That the excesses of the 'war on terror' were undertaken under the banner of spreading democracy only seems to have enhanced the growing reputation in the developing world of Western liberal democracies – the foremost promoters of democratisation around the globe – as self-interested and hypocritical. Their flat-out rejection of the first truly open and free elections in the Arab world, which took place in Palestine in 2006 and brought Hamas to power, will not help repair this reputation. It is difficult to expect a global surge in enthusiasm for liberal democracy when its chief sponsors set a less-than-perfect example.

In the meantime, a new wave of securitisation appears to be sweeping the globe. In the West, civil liberties are being curtailed in the interest of national security, and in many developing countries the demands of national security are being used to legitimise the authoritarian tendencies of the state, and to postpone indefinitely the projects of reforming and democratising the military institution. 'Appalled by the mess next door', reads a recent article in *The Economist*, 'few Syrians now doubt that their own secular dictatorship is preferable to the anarchy of supposedly democratic Iraq.'[30]

Add to this the spectacular rise of a non-democratic China, an increasingly aggressive and authoritarian Russia, and the failed attempts at democratic revolution in Ukraine and Lebanon, and the short-term global prospects for democratisation and military reform look less promising than they did in the early 1990s.

State

In Max Weber's classic definition, the state is a set of administrative institutions that holds 'the monopoly of the legitimate use of physical force within a given territory'.[31] These institutions include the military, legislative, executive and judicial bodies, the civil bureaucracy and the police. In this section, the state-institutional influences on the civil–military balance, and the evolution of the relevant institutions through reform initiatives, are analysed. The role that individual actors – politicians, generals, etc. – play in leading or resisting reform initiatives is also examined. In short, this section is concerned with the *processes*, and counter-processes, of military reform and democratisation at the state level, in light of the social and global influences outlined in the previous sections.

How can such processes be analysed and performance be measured? The approach of SSR is useful in studying reform initiatives at the state level, as it is chiefly concerned with monitoring reform and transition, and recommending policy to decision-makers to achieve the 'accountability, oversight and transparency' of the security sector.[32] One aspect of reform that SSR puts particular emphasis on is the issue of consolidation. While jump-starting military reform is vital to democratising a country, maintaining and strengthening reforms can prove trickier. For instance, although in central and eastern Europe, as Edmunds notes, 'armed forces have been removed from domestic politics … in most countries [in the region] the military still has a disproportionate (and sometimes exclusive) influence over defence policy'.[33]

In many democratising countries, removing the military from politics has proven a simpler task than keeping it away from politics. Ensuring the military establishment accepts its new, less influential position is possibly the biggest challenge of military reform. The consolidation of reforms may prove difficult when middle- and lower-ranking officers who possess well-developed political and ideological views despise their new position and loss of power, and 'feel threatened by the forces active and dominant in the new democratic politics'.[34]

Powerful military establishments may attempt to expand their influence into the post-transition period through, for example, trying to ensure irreversible constitutional change (as was the case in Turkey post-1983), assuming key positions in the new government, or working to preserve the future autonomy of their own personnel and finances (both of which have been the case in Indonesia post-1998).[35]

Consequently, SSR divides reform into two categories: 'first generation reforms' are concerned with 'the establishment of appropriate structures for (democratic) civilian control, and measures to depoliticise security sector actors and remove them from partisan intervention in domestic politics', while 'second generation reforms' deal with 'the further consolidation of democratic procedures of oversight and transparency, the way structures and institutions implement policy, improvements in effectiveness and efficiency, and the wider engagement of civil-society'.[36]

Edmunds lists three methods of measuring performance across the stages of security-sector reform. First is the 'generic framework', which provides a 'normative, "ideal type" against which performance can be measured'. Second is the 'collective/regional approach', which measures performance 'against specific international institutional agendas', such as NATO or EU criteria. A third measure is the 'process/facilitation approach', which focuses on specific empirical, as opposed to normative, criteria, such as measuring transparency or oversight, rather than 'democracy' in the abstract.[37]

This paper will draw in particular on the first two measures outlined above as it considers the success and shortcomings of Turkey's and Indonesia's democratisation and military-reform efforts, and examines the specific internal and external factors that influence these efforts.

CHAPTER TWO

Turkey

Society

The military's perception of its role in Turkish society: 'guardians' of the secular regime

Ironically – or perhaps not – Turkey's journey of modernisation and Westernisation began with the military, when the Ottoman Sultan Mahmut II abolished the Janissary corps to establish a modern, Western-style army in 1826. From then on, the military became 'an avenue of contact with the West'.[1] It was mainly through the military that Western ideas such as nationalism and secularism were introduced into Turkish society. Both the Young Turk revolution of 1908, which initiated a short-lived constitutional era, and the Turkish nationalist movement led by Mustafa Kemal Atatürk after the First World War, which led to the foundation of the modern Turkish Republic in 1923, were undertaken by Western-oriented military officers.

Despite fighting occupation by the West, Atatürk and his fellow nationalist officers envisioned the new republic to be 'Western', which they equated with progress and modernity. Breaking from the multi-ethnic and Islamic heritage that had been at the centre of Ottoman state and society – and that was largely blamed for the empire's backwardness and collapse – the new Turkey was to be a secular nation-state.

The modern Turkish armed forces (the acronym of their title in Turkish is TSK) perceive themselves to be the heirs of these founders of the secular republic, and consequently ascribe to themselves the role of 'guardians' of the regime's core values and interests. These values and interests are embodied in Kemalism, the official ideology of the state, which has been described as 'a project of politically constructing and manipulating a modern Turkish nation-state on secular and western rather than Islamic precepts', which relies 'on the officer corps as the main carrier of [its] positivist-progressive ideals'.[2]

The TSK views its guardianship role not only in terms of defending the country against armed aggression, but as making it an active and legitimate actor – arguably the most important one – in the affairs of the state. Military analyst Mevlüt Bozdemir suggests that underpinning the guardianship concept is a kind of 'elite revolutionism', or 'Jacobinism', driven both by the military's self-confidence, stemming from its role in founding and constructing the republic, and its lack of confidence in the 'other', namely the uncontrolled masses, the population at large.[3] Therefore, what 'guardianship' really stands for is the military's desire to supervise, and ultimately to have a monopoly of control over the fate of the Turkish people.

In 1923, in a country ruined by decades of war and destruction, which had no bourgeoisie to revive the economy, and which had turned its back on its social and political heritage, the military had indeed stood as the only organised institution capable of constructing a new state, and a new national identity. In the early decades of the republic, the military – or individuals with a military background – filled the boots of politicians, bureaucrats and businessmen with considerable success.

But when Turkey eventually emerged from its fragile post-war condition to become a stable regional power, the military's role as guardian of the secular regime remained unquestioned. Since Turkey's first multiparty elections in 1946, the TSK has ousted four democratically elected governments – in 1960, 1971, 1980 and 1997 – each time legitimising the intervention in terms of its duty to 'guard' the regime.

Throughout their professional lives, from the day they enter the military academy onwards, Turkey's military officers are reminded of their mission to guard and guide the republic in accordance with Kemalist principles. The fundamental elements of this mission can be defined as protecting national unity and upholding the authority of the secular and centralised state apparatus. According to political scientist Tanel Demirel, the 'state' in the traditional Turkish military context is not understood in the classic liberal sense, as a mechanism to ensure that individual liberties are safeguarded based on a social contract, but rather as an 'abstract entity with autonomous rights over society'.[4] And while the line between the state and the people is quite clear, the role of the military within the state remains blurred. In an indirect way, the military – with its supra-political position and ability to influence other institutions of the state with limited account-ability to the people – *is* the state in Turkey. In this context, being a soldier is not considered just as doing a professional job, but rather as performing a sacred duty towards the state.

A new type of guardianship: the national security regime

During the Cold War, communism was perceived to be the main threat, both internal and external, to the Kemalist regime. The end of the Cold War and the dissolution of the Soviet Union and Yugoslavia brought forth new security issues in Turkey's immediate neighbourhood, from the Balkans to the Caucasus and the Middle East. At the same time, globalisation and economic liberalisation started to undermine the sovereignty of the nation-state, as well as the monopoly of control of the Kemalist elite over Turkish public opinion. In the 1990s, faced with widespread challenges to its ideological monopoly over society, and equipped with unprecedented constitutional authority thanks to the 1980 coup, the military's perception of its guardianship of the regime evolved into what is commonly referred to as the 'national security regime'.

Turkish scholar and editor Ahmet İnsel has observed that the national security regime is founded on the notion that there are constant and imminent threats to the regime, and that Turkey is plagued by internal and external enemies.[5] According to this view, Turkey's unique geopoliti-cal position – between East and West, and in the midst of conflicting and 'dangerous' ideologies – makes it a country whose very existence is under ceaseless attack. Externally, all neighbours, as well as the West, are viewed as potential enemies that would like to see a weaker Turkey.[6] Internally, the threats are various, although political Islam and ethnic separatism – threat-ening secularism and national unity, the two sacred pillars of Kemalism, respectively – are considered the main enemies of the state. The fear of ever-present enemies bent on carving up Turkey gives the military the mandate it needs to stay involved in Turkey's political affairs.[7]

The national security regime involves the arrangement of political and social issues into a careful hierarchy. On top are the 'national secu-rity' issues, which directly concern or threaten the regime. The definition of these threats, as well as the parameters of public discussion of them, is set by the state, i.e., the military and the bureaucratic elite, in accor-dance with the official ideology. Below are the 'political' issues, which are considered less urgent and less threatening matters, and include most economic, social or cultural issues. The discussion of a 'political' issue may be conducted publicly, so long as it does not clash with a national security issue, in which case the latter has absolute priority, and the state the last word. For example, religious freedom is a political issue as long as it does not involve a critique of secularism, or the state's definition of secularism. Similarly, if discussing cultural identities within Turkey brings 'national unity' into question, then this becomes a national security issue.

While the classic notion of military guardianship defines the principles that it is the military's duty to protect – namely, state authority, national unity and secularism – the national security regime places those principles directly within a threat framework. In this way, the military and the secular elite attempt to establish a monopoly on defining threats.[8] This 'insecure security' ideology, whose continuing legitimacy depends on the perpetual existence of the threats it fights against, is a means for the state as embodied by the military and the civil-service elite to maintain its control over the fate of the regime and society.

The national security regime relegates democratisation to the status of a low-urgency issue at best. In fact, the 'threat and fear regime', as analyst İlhan Uzgel calls it, constitutes a formidable obstacle in itself to the advancement of democracy and human rights in Turkey.[9]

Societal perceptions of the military: guardianship unquestioned

The enduring popularity of the TSK within Turkey never fails to puzzle outside observers. It might seem unlikely that a military establishment which sees itself as the patent-holder of the state, which has repeatedly intervened in civilian politics and which has a deep lack of trust in 'the people' is still the institution that is the most trusted by the people. Yet, year after year, the military comes out in national surveys as the most popular and trusted institution in Turkey.[10]

Part of the explanation is cultural. A deeply embedded view that has its roots in religion and history considers soldiering an integral part of the collective identity of the nation.[11] The saying 'every Turk is born a soldier' sums up this still-popular belief in the 'army nation'. Even today, men are often sent off for their mandatory military service amid jubilant celebrations, and those who do not serve are regarded with scepticism and even disrespect. Although to those with better education and brighter financial prospects military service tends to appear more as a burden, for the greater part of the population, it remains a sacred duty and a necessary step towards attaining manhood.

The military is also popular because it is widely perceived as being meritocratic, successful, altruistic and incorruptible. Academic Ahmet Taner Kışlalı has observed that the majority of officers in the Turkish military come from lower- and middle-income families.[12] This supports the broadly accurate belief held by many Turkish people that the TSK is made up of the nation's best and brightest minds, regardless of financial background or personal connections.

As for success, the TSK is generally thought to be the only highly successful institution in Turkey. From the Battle of Gallipoli in 1916 to the founding of the secular republic, from Cyprus in 1974 to the fight against Kurdish separatists in the 1990s, the military has played the lead role in most of the success stories that are a source of national pride in modern Turkish history. It is also a well-organised and highly professional body that operates efficiently. Kemal Dervis, the current head of the United Nations Development Programme, who left the World Bank in 2001 to become Turkey's finance minister, has said that 'only in the military one comes across a successful [sic] team work'.[13]

The image that the military enjoys of altruism and incorruptibility is mirrored by the widespread public conviction that politicians, businessmen and the media – the main components of civil society – are self-interested, corrupt people who cannot be trusted. This conviction, not wholly baseless, has been successfully exploited by the TSK to justify its political interventions.[14]

Finally, the military is the most trusted institution in Turkey because Turks are rarely taught otherwise at school or in society. On mandatory 'national security' courses in high school, and during compulsory military service, young Turks learn about the sacredness of military service and the unquestionable position of the TSK in society.[15] Commentator Ali Bayramoğlu has noted that despite holding the military in as high esteem as they do, most Turkish people actually know little about the institution.[16] The TSK is highly secretive, and to criticise it publicly is still taboo. Those who do so run the risk of being labelled unpatriotic – not necessarily by the military, but by 'the people' themselves.

Military and societal perceptions of democratisation and military reform

When the military toppled the civilian government on 12 September 1980, Kenan Evren, the junta leader, announced that the TSK had the 'historic task' of restoring 'the authority of the state' and ensuring 'the re-establishment of a functioning democracy'.[17] After every direct intervention in politics, the military in Turkey has vowed to 'restore democracy' after it has been abused by politicians. This leads one to the question of how the military defines democracy, this system which civilians regularly abuse and which soldiers have to restore.

According to political analyst Metin Heper, what the TSK calls democracy is in fact the ideal of a 'rational' or selective system of government: 'an activity among the knowledgeable and patriotic persons who try to find the best policy and thus promote the general interest'.[18] The military thus excludes those civilians whom it deems neither knowledgeable nor patriotic enough to participate in political activity. When criticised for its regular interventions in politics, the TSK usually responds that the military will get involved until Turkish democracy reaches a certain level of maturity.[19] Coexisting with the national security regime, this attitude creates a fundamental paradox, since democracy cannot mature while important and sensitive socio-political issues remain taboo.

Perhaps surprisingly, a sizeable portion of the Turkish population seems to accept this notion of 'rational democracy'.[20] From a minimalist perspective, Turkey has been a multi-party democracy since 1946. Turkish people frequently demonstrate their faith in the ballot box with high voter turnout, exercising their right to reward or punish political parties. From a substantive point of view, however, one can argue that Turkish democracy remains unconsolidated. Linz and Stepan define a democracy as consolidated when 'a strong majority of the public opinion, even in the midst of major economic problems and deep dissatisfaction with incumbents, holds the belief that democratic procedures and institutions are the most appropriate way to govern collective life'.[21] Public support for regular military interventions in politics, and occasional public appeals to the generals to protect the nation's unity and integrity against various treacherous agents, suggest that many Turks still have more faith in the institutions of the state as currently constituted than they do in democratic institutions.

In 2004, analysts Mark Tessler and Ebru Altınoğlu concluded an extensive survey of Turkish 'attitudes towards democracy, the military and Islam' with the following observation:

> A majority of Turkish citizens do express support for democracy as a political system, but ... at least some of these individuals are probably expressing satisfaction with the status quo rather than a genuine commitment to democracy. Even more important, both the limited importance attached to political liberty and the fact that importance attached to political liberty is not related to

support for democracy suggests that Turkey does not at present possess a democratic political culture.[22]

They further noted that 'the anomaly of an anti-democratic institution presenting itself as the guardian of democracy may neither trouble most Turkish citizens nor be taken especially seriously'.[23] In fact, most Turks feel there is nothing strange about the military's involvement in politics. The high school national security courses mentioned above, which are taught by military officers, and which are the only classes in which the discussion of current affairs is permitted, establish the impression early on in Turkish citizens' lives that politics is a primarily military activity.[24]

This understanding of the military's political role shapes social attitudes towards military reform and democratisation in Turkey. In addition, even if Turks were convinced of the importance of military reform, for a great part of the population, such issues are considered trivial at best. 'In a country where poverty is widespread, in which most people can hardly survive on daily basis, it is impossible to think of civil–military relations as an urgent matter', writes Demirel.[25] 'Instead, it is regarded as a problem that only the EU and some intellectuals care about, which has little bearing on Turkey's real agenda.'[26] At worst, it is seen as part of a sinister scheme by Turkey's internal and external enemies to weaken the Kemalist regime. Both perceptions make it especially difficult for Turkey's least trusted people – politicians – to criticise and reform Turkey's most trusted institution.

Under the national security regime, Turkish people are typically presented with a choice between extreme conditions. As journalist Ece Temelkuran notes, a favourite topic of discussion for middle-class Turks is whether one would prefer a military coup over the imposition of sharia (Islamic) law, or vice versa.[27] Convinced that Islamic fundamentalism poses an imminent threat to their lifestyles, most secular Turks have traditionally welcomed the coup. Only in recent years, with EU-driven reforms and the influence of globalisation over society, have some Turks come to realise that there is a third option: democracy.

That at least some demonstrators chanted 'no to sharia, no to coup; yes to [a] democratic Turkey' in the pro-secular, pro-military rallies of spring 2007, and that some voters expressed their objection to the use of undemocratic means to block the political process in the July 2007 parliamentary elections, suggests that attitudes may be changing, and that belief in this third option – though still held by a minority – is slowly gaining momentum among the people of Turkey.

The global order

The Turkish military and the global order

To fairly analyse Turkey's military reforms and democratisation, observers must look at Turkey in its geopolitical context. No democratic European country that has expressed discontent with the slow pace of Turkish reforms has neighbours as challenging as Turkey's: an increasingly assertive and authoritarian Russia to the north; the corrupt and confrontational regimes of the Caucasus to the east; and a dictatorial Syria, a chaotic Iraq and an ambitious, theocratic Iran to the south. As international and strategic relations analyst Ali Karaosmanoğlu has noted, 'Turkey needs a peaceful environment and secure borders for an effective and orderly continuation of the reform process'.[28] Pursuing military reforms and democratisation with vigour while simultaneously battling ethnic separatism and fending off regional instabilities is no easy task for any country.

That being said, the Turkish military's perception of the global order, as embodied in the national security regime, exploits the geopolitical challenges facing Turkey, and does not reflect the realities of the new strategic landscape created by the globalisation of security affairs in the post-Cold War era. Virtually everywhere in the world aside from Northern Cyprus and perhaps Azerbaijan is viewed by the TSK as a threat to Turkey's national security. This author remembers walking into a national security lecture in high school and seeing the entire blackboard filled with country names, with the title at the top reading: 'Our Enemies'. As late as 2005, due to disputes over territorial waters and Cyprus, Greece, which has a population one-fifth the size of Turkey's, figured as an imminent threat in the strategic assessments of the TSK.[29] This challenging attitude – illustrated by the fact that between 1994 and 1998 Turkey ranked third globally in arms purchases – most likely contributes to Greece considering Ankara a significant threat to its national security in turn.[30] The military uses the supposed Greek threat, among others, to legitimise defence spending, which, along with education, is the country's highest single budgetary expenditure.[31]

The globalisation of security affairs may be characterised in terms of nations going beyond the traditional national security framework, implying 'multilateralism, interdependence and a shift from active confrontation to cooperative arrangements'.[32] Political scientist Ümit Cizre argues that Turkey has failed to adapt to these changes in the post-Cold War era:

Rather, the systematic reinvention of Cold War-era security concepts helped to restore the *status quo* and (re)legitimize the Turkish armed forces as the guardians of the regime. The securitization of every aspect of life was prioritized and Turkey failed to embrace a security approach that safeguarded human rights and enhanced an understanding of the democratic character of the state.[33]

Even so, the TSK's approach to foreign affairs cannot be summed up as simply paranoid and confrontational. The military seems to be in a complex position in its dealings with the wider world; shifting between cooperation and power politics. As Karaosmanoğlu suggests:

While its EU candidacy, NATO membership … and participation in peace operations are inspiring internationalization, multilateralism, cooperative security, democratic control of the armed forces, and emphasis on societal and individual security, [Turkey's] regional environment is suggesting security through power politics and the sustained primacy of the nation-state.[34]

An ambivalent posture is especially evident in the military's approach to Western and international organisations. On the one hand, the TSK has been highly cooperative with NATO and the UN. In the 1990s, the Turkish military participated in the UN Protection Force, as well as several other NATO and UN missions in Bosnia and Kosovo. It assumed the command of the UN operation in Somalia and twice that of the International Security Assistance Force in Afghanistan.[35] Turkey actively contributes to the UN's Peace Support Operations and to NATO's Partnership for Peace programmes.[36]

In Europe, Turkey participates in the Organisation on Security and Cooperation in Europe, and, although not an EU member, it has informed the EU of its military's readiness to contribute to the forces planned under the European Security and Defence Policy. Regarding Turkey's EU membership bid, TSK generals have commonly made supportive remarks. General Hüseyin Kıvrıkoğlu, chief of the general staff in the 1990s, declared that 'joining the EU is a geopolitical necessity', while General Yaşar Büyükanıt, the current chief, has said that 'it is Atatürk's path and we cannot be against the EU'.[37]

On the other hand, there has in recent years been growing suspicion within the Turkish military of both the EU and the US. This suspicion, echoed in the pro-military rallies of spring 2007 in the slogan 'no to the EU, no to the US', seems to be most prevalent among retired officers and hard-liners. In 2005, Navy Commander Admiral Yener Karahanoğlu commented: 'I believe we will have to count our fingers after shaking hands with the EU'.[38] In a recent conference of the leading think tank of the secular elite, the Kemalist Thinking Association, retired General Tuncer Kılınç, former secretary-general of the National Security Council (a powerful national security advisory board made of up of top civilian and military represen-tatives, in which until recently the generals had the upper hand) argued that Turkey should quit NATO 'in order to free itself from Western hege-mony and exploitation', and form an alliance with Russia, China, India and Iran.[39]

The military's wariness of the EU is due to the strategic demands that the latter makes of Turkey, and the increasingly confrontational attitude shown by some European countries towards the country. Moves by several European nations to recognise the massacres of Ottoman Armenians in 1915 as genocide and the EU's demand for Turkey to recognise the govern-ment of Cyprus are seen as signs of European hostility, and many Turks also believe that Western Europe is the financial base of the separatist Kurdistan Workers' Party (PKK), a major thorn in Ankara's side.[40]

Regarding the US, the military is unhappy with Washington's coopera-tion with the current moderate Islamist Turkish government, and with its support for the Iraqi Kurds, which is perceived to be the root cause of the revival of separatist activities in eastern and southeastern Turkey. A 2003 incident in Northern Iraq, in which US forces arrested and 'hooded' 11 Turkish special operatives, infuriated the Turkish public and caused anti-American sentiments to rise sharply. Most recently, the passage in October 2007 of a resolution in the US House Foreign Relations Committee calling the Armenian massacres of 1915 genocide, and Washington's strong oppo-sition to a Turkish incursion into Northern Iraq to root out PKK bases there, are likely to add to the increasingly unfavourable image of the US in Ankara.[41]

For the most part, however, it is the profoundly suspicious mindset of the Turkish state that lies behind the current anti-Western reaction. In a rapidly changing global atmosphere, this mindset urges Turks who wish to preserve the status quo to see the entire world, and especially the West, as bent on weakening Turkey's secular and centralised state structure.

The military and the secular elite, in using the National Security Council and their influence over public opinion to securitise the principal foreign affairs issues and elevate them from the realm of the political to the realm of national security, not only limit the depth of public debate of these issues, but also cause Turkish society to perceive international affairs with a military mindset, interpreting the global order in black-and-white terms of friends and foes.

What Turkish military reforms need to achieve is to ensure that foreign policymaking is a civilian, rather than a military, activity. This will entail both military and civilian leaders adopting a more self-confident and open-minded approach to globalisation that is capable of challenging the national security regime's view of Turkey in the world.

The EU as the main anchor of Turkish democratisation

Turkey has been an associate member of the Western European Union since 1963, and has been striving for full membership in what was then the European Community since 1987. In December 1999, the EU finally accepted Turkey as a candidate country, and in October 2004 it began membership negotiations with Ankara. In 1999, upon accepting it as a candidate country, the EU offered Turkey a 'pre-accession strategy', in the form of 'harmonisation packages', 'designed to stimulate and support its reforms'.[42] Fuelled by overwhelming public support for the EU bid in Turkey, these packages led to the implementation of reforms between 2001 and 2004 that are considered to have been breakthroughs for Turkey's democratisation process. A particular achievement was the restructuring of the military high command in 2003 as part of the seventh harmonisation package, which has come to be praised by Turkey's liberal intellectuals as the 'little revolution'.[43]

How do we explain the (once) popular support for EU membership in the context of widespread indifference, or even hostility, towards military reform and substantive democratisation? The answer lies primarily in the fact that not all Turks support the EU for its democratising role. A nationwide poll taken in 2002 revealed improvement in economic welfare as a result of EU membership as the main reason given by many Turks for favouring Turkey's bid.[44] That was the year in which support for the EU peaked in Turkey;[45] it was also the year when Turkey was struggling to recover from two devastating economic crises, with unemployment and inflation pushing record highs.[46] Membership in the EU was seen as a solution to the country's chronic political and economic instability, for which the corruption and mismanagement of inefficient coalition governments were widely held responsible. Five years on, with the economy considerably recovered[47] thanks to the stability provided by a single-party government and relatively sound economic policies, Turkey is once again plunged into a tense political atmosphere in which threats to national unity and secularism, rather than the economy, are making the headlines. In line with this change, support for EU membership declined from 67% in 2004 to a mere 32% (and shrinking) in 2007.[48]

In its 2005 progress report, the European Commission openly stated its expectations of Turkey's military reforms: 'Turkey should work towards greater accountability and transparency in the conduct of security affairs in line with member states' best practice. In particular, statements by the military should only concern the military, defence and security matters and should only be made under the authority of the government.'[49] More recently, after the TSK made a statement on 27 April 2007, known as the 'e-memorandum', confronting the AKP government on issues of secularism and the presidential election, Commissioner Olli Rehn condemned the intervention and asked the Turkish military to leave the remit of democracy to the democratically elected government.[50]

In essence, Brussels has one rather straightforward political demand of Ankara: democratisation. It has made it abundantly clear that it is impossible for Turkey to join the club unless civil–military relations are reformed to meet the EU's democratic standards. The EU's resulting influence in this area has been substantial: as an overview of Turkey's military and democratic reform initiatives will demonstrate, over the past decade, Turkey has made no serious democratic stride without the EU's push.

The EU has become a role model against which Turkey measures the goals and outcomes of its democratisation process. Huntington wrote in 1990 that 'the EC can be an anchor for Turkish democratisation', and so it has been for the 17 years since.[51] The precedents of other countries would suggest that this anchor could indeed enable a country like Turkey to democratise: prior to becoming EU members, both Spain and Greece had had more experience of military dictatorship, and less of democracy, than Turkey has currently. By the same token, however, 'failure to provide that anchor would make the future of Turkish democracy more uncertain'.[52] Heper argued in 2002 that 'Turkey's present aspiration to become a full member of the European Union will further discourage the military to make overt interventions in politics'.[53] But five years later, we observe that Turkey's current lack of aspiration to join the EU encourages the military to interfere more actively than previously in civilian affairs.

In December 2006, Brussels partly suspended negotiations with Ankara following a row over Cyprus, one of the most sensitive issues covered by the accession talks. Turkey and the EU have not fully turned their backs on each other, but the relationship is certainly growing cooler.

There are a number of factors on both sides that might explain the downturn in relations. We have already seen some of the reasons behind the loss of public enthusiasm for the EU in Turkey, and the fears and resentments of the country's secular establishment. Perhaps the most crucial issue on the EU side is the growing lack of popular support for further enlargement and especially for Turkish membership. This opposition found its voice in the French and Dutch rejections of the EU constitutional treaty in 2005, and the recent changing of the political guard in a number of European countries has seen the replacement of pro-Turkey politicians with leaders openly against Turkish membership, namely Angela Merkel in Germany and Nicolas Sarkozy in France. In the meantime, Turkey's bilateral relations with several European countries, France in particular, have been deteriorating over the issue of recognising the massacres of Ottoman Armenians as genocide.

Strategic arguments in favour of a democratic Turkey tied strongly to Europe – that it would keep at bay a 'clash of civilisations' between the West and the Muslim world; that it would expand Europe's politico-economic influence further towards the Black Sea region and the Middle East; secure vital energy routes to counter the Russian energy monopoly; and inject dynamism into stagnant European economies – fall on deaf ears in Europe, where the public seems to be more concerned with Turkey's Muslim identity, its historical and cultural legacy on the continent, and potential immigration issues.[54] This reinforces the perception among Turks that the EU is a 'Christian club' that will never admit Turkey, even if Ankara put in place all the required reforms.

The EU's recent success in Central and Eastern Europe has been attributed to the Union's 'positive conditionality' principle, which gave candidate countries a realistic hope that their efforts would pay off in the foreseeable future. Without such hope or encouragement, Turkey cannot be expected to continue on the EU track forever, and, if derailed from this track, it risks losing the main engine that has fuelled its domestic reform initiatives until now.

The influence of NATO and the US on Turkish military reform and democratisation

The influence of NATO on the Turkish military has been largely modernising, rather than democratising.[55] When Turkey decided to join this strategic alliance at the end of the Second World War, it undertook serious structural reforms to bring the military on par with Western standards.[56] Since 1953, Turkey has received over €4.5 billion in funds from NATO, almost all of which has been used for upgrading military equipment and facilities.[57]

The US role in Turkish democratisation, and Washington's evolving relationship with the TSK, are more complex. Stephen Larrabee and Ian Lesser of RAND argue that 'support for Turkish democratization and human rights has been a consistent theme of American policy, with successive American officials urging Ankara to "take risks for reform"'.[58] Yet history suggests that Washington's deeds have not always matched its words.

During the Cold War, the US had a strategic relationship with Turkey, based on cooperation within NATO against the Soviet and communist threat. In return for Turkey's assistance with the Korean War effort and for its willingness to host US weapons on its soil at the height of the nuclear arms race with the Soviet Union, Washington generously supported Turkey, both militarily (through NATO) and financially (through the Marshall Fund). This also meant, however, turning an occasional blind eye to undemocratic impulses within the country. In the autumn of 1980, for example, troubled by the revolution in Iran and the Soviet invasion of Afghanistan, Washington covertly helped engineer the military coup in Turkey.[59] In its strategic relationship, the US appeared to value Turkish stability over democracy.

Although the end of the Cold War brought the indispensability of this strategic partnership into question, the US–Turkish alliance survived the 1990s, as the two countries' militaries cooperated in the Gulf War, as well as in a number of UN and NATO missions in Somalia and the former Yugoslavia.[60] The relationship began to crack, however, when, in March 2003, the Turkish parliament rejected a proposal to allow US forces to open a northern front against Iraq on Turkish territory. This was followed by the notorious 'hooding' incident in Northern Iraq, and relations between Turkey and the US reached an all-time low.[61]

The TSK had been in favour of the March 2003 proposal. However, it had on this occasion refrained from influencing the Turkish parliament's decision. When the parliament, led by the AKP government, rejected the proposal, Paul Wolfowitz, then US deputy secretary of defense, criticised the Turkish military for 'not fulfilling its leadership role', implying that the US would rather have the military intervene in Turkey's civilian politics in those cases where it was in Washington's interest.[62]

Washington's current position towards Turkey, and actors within Turkey, can best be defined as ambivalent. Initially, the Bush administration seemed to find a new ally in the moderately Islamist AKP; Washington saw a pro-American Muslim democracy in the Middle East, and a NATO ally in the EU, as in America's strategic interest. Consequently, the US has supported Turkey's EU bid, occasionally urging European countries to adopt a more constructive approach towards Turkey, and commending the AKP government on its drive for reform.[63] These expressions of support seem to have provided the AKP with a sense of security and confidence in implementing sensitive military reforms, especially in its first two years in government, since, its leaders must have thought, Turkey's generals would think twice before toppling a government openly backed by the US.

However, the war in Iraq took Washington's attention away from Turkey's reform initiatives. The Turkish government's role in rejecting the March 2003 proposal, along with Prime Minister Tayyip Erdoğan's confrontational approach towards Israel, Turkey's rapprochement with Syria and Iran, and the AKP's willingness to recognise the Islamist Hamas government in Palestine seriously strained the alliance between the AKP and the White House.[64] At the same time, the growing instability and chaos in Turkey's immediate neighbourhood as a result of the war has mostly worked to divert the Turkish government's energy and the public's focus away from the issue of democratisation.

In addition, Washington has been unwilling to sacrifice its relationship with the Turkish military, a permanent actor and a force to be reckoned with in the region, as well as a key NATO ally, for its alliance with the AKP government. But US support for the Iraqi Kurds – America's sole whole-hearted helpers in Iraq – has put Washington in a difficult position in this regard, as Ankara and the Kurdish administration in Northern Iraq are in dispute over a range of potentially explosive issues, including Turkey's Kurdish question and the fate of the oil-rich city of Kirkuk in Northern Iraq.[65]

America's conflicting interests in relation to Turkey could be seen in Washington's response to the TSK's e-memorandum of April 2007: the White House neither condemned nor approved the military's blatant intervention in politics; Assistant Secretary of State Dan Fried stated instead that the US wished to remain neutral on the issue.[66]

On the whole, it is difficult to suggest that American interests in Turkey have had a boosting influence on the country's democratisation process: it is in the interest of the US for Turkey to be a liberal democracy, at least formally, but it is also in its interest to keep Turkey as a political and economic client state on whose strategic cooperation Washington can continually rely. Recent history demonstrates that the US often falls short of supporting substantive democratisation in Turkey when this implies a potential divergence of strategic interests between Washington and Ankara. This ultimately enables Turkey to remain a mediocre democracy at best.

State

Military interventions and the creation of a 'military democracy'
It would be misleading to look at Turkish democratisation from a binary perspective, juxtaposing an authoritarian military against democratically inclined civilian politicians. Turkish democratisation is best studied through a holistic approach that analyses the evolution of democratic political culture in Turkey.

It should be recognised that the military, or individuals with a military background, were behind most of Turkey's early strides towards democracy. Turkey became a multi-party democracy when the Kemalist elite saw fit, not when the pressure from the people became insurmountable. Likewise, Turkey's most democratic constitution to date, the 1961 constitution – which granted unprecedented rights to freedom of thought and expression, and promised a 'social state' – followed from the military coup of 1960.[67]

According to Cizre, the military's acceptance of the legitimacy of civilian rule, as evidenced by there being a return to civilian rule following almost every intervention, 'distinguishes it from armies elsewhere in the Third World'.[68] Furthermore, the Turkish military refrains from being associated with any political party, or any civilian or military leader (except for Atatürk), and carefully garners popular support – with the help of its allies in the media, politics and civil society – before an intervention.

The Turkish military cannot be held responsible for all of Turkey's democratic shortcomings; civilians, too, share part of the burden. That being said, each of the military's interventions, whatever its cause or purpose, has led to grave human-rights abuses, as well as to constitutional changes that have seen the military's political, legal and economic authority and autonomy expand. This has invariably worked against democracy in Turkey, which is still struggling to free itself from the grip of the semi-authoritarian system established after the 1980 coup.

Constitutional change undertaken by the military is a well-established political habit in Turkey. Following the 1971 intervention, the 1961 constitution was heavily amended. The most important change to tilt the civil–military balance in favour of the military was the establishment of the National Security Council, which the TSK introduced into the 1961 constitution as, in Cizre's words, 'an embodiment of the bureaucracy's primacy over the popularly elected parliament'.[69]

Amendments in 1973 and a new constitution in 1982 following the 1980 coup turned the National Security Council into something like a government above the government, which exerted tremendous influence over the elected government and the parliament, and which was not accountable to any elected policymaker.[70] In 1982, the number of military members of the council was increased at the expense of civilian members. Its general secretary, a military man, was given extensive powers, and the council's recommendations to the council of ministers (which has a similar function to the British Cabinet) were given 'special consideration', making them tantamount to official edicts.[71]

Through its influence over the National Security Council, the military was able to gain control of the drafting of the National Security Policy Document, a secret document, updated every few years, that identifies internal and external threats to national security and lays out the state's national security policy.[72] The document, which has been dubbed Turkey's 'secret constitution', was once referred to by former Chief of Staff General Doğan Güreş as 'the god of all policies, the mother of all laws: it is unthinkable to act against it'.[73]

The 1982 constitution also redefined national security in such a way that it could be interpreted to cover any policy field.[74] Furthermore, it gave the National Security Council the right to govern through 'emergency rule' – martial law, in effect – which was subsequently used as the basic instrument of managing the Kurdish problem in the 1990s.[75] The constitution also established two regulatory bodies, the Higher Council of Radio and Television and the Council of Higher Education, designed to ensure effective state control over media broadcasts and university policies and curricula respectively. Both institutions would have National Security Council representatives in their boards, and their directors were appointed by the president.

After 1982, the presidency, formerly a symbolic post, was equipped with wider authority and the military made sure that future presidents would either be former generals, or civilians approved by the National Security Council.[76] Finally, the constitution continued what the 1973 amendments had begun, and made the TSK's budget and finances, as well as the judiciary, virtually exempt from civilian oversight.

According to scholars Ilkay Sunar and Sabri Sayari, the 1982 constitution envisioned 'a state divorced from politics and a depoliticised society'.[77] It not only expanded the military's authority over and independence from civilian politics and public life; it also took away most of the democratic rights and freedoms granted by the 1961 constitution. After the period of direct military rule ended in 1983, the TSK was able to consolidate its position thanks to the 631 laws it had enacted since 1980 that could not be changed or criticised.[78] Turkey became a military democracy; 'a democracy without freedoms'.[79]

While the military was able to substantially expand its influence over politics through the 1982 constitution, opposition to the junta within society had grown over the course of its three years in government (the longest period of direct military rule there has been in Turkey). In the liberal economic era that followed, this opposition gave rise to a culture of careful criticism of the military's involvement in politics. What prevented this trend from developing any further was the popular conviction, which the military helped to shape in the early 1990s, that political Islam and Kurdish separatism presented imminent threats to Turkey's national security, with which only the TSK was capable of dealing. Political parties, democratic pressure groups and the media became cautious, and criticism of the military's regular meddling in politics diminished. In 1997, in a confirmation of its political authority and the rewards of its indirect control over the state and civil institutions, the military, via a host of intermediaries in the parliament, NGOs and the media, forced the Islamist government of Necmettin Erbakan to resign without having to fire a bullet. Since labelled the 'post-modern coup' because it was achieved through the use of influence, rather than as a physical takeover, this intervention demonstrated just how extensive was the TSK's reach into all aspects of society.

Military reform and democratisation following the 1999 Helsinki Summit

The crucial boost to Turkey's democratisation process came only two years after this low point, in December 1999 at the Helsinki Summit, when the EU presented Ankara with the pre-accession strategy intended to encourage it in its reforms.[80]

Political parties, seeing the EU bid as the key to winning elections, promised reforms to bring the country up to European standards. The boldest reforms aimed at democratising and readjusting the civil–military balance were undertaken after the November 2002 general election, when the Turkish electorate punished the traditional parties that had come to represent the corruption and political stalemates of the previous decade by giving the moderately Islamist AKP a mandate to form the government alone.

The AKP is the 'modernist' faction of the Virtue Party, which was the last in a series of Islamist parties, including Erbakan's Welfare Party, that succeeded one another upon being shut down by the state for being anti-secular. Breaking from the identity politics and the strict Islamic values of their predecessors, the AKP's leaders advocated a 'politics of service' based on liberal economic values, thereby moving their party from the margins of the political spectrum to the mainstream. The AKP at this point not only represented the political evolution of Islam in Turkey, but also the rise of a new middle class whose roots were in the poor, rural and conservative countryside that had typically been looked down upon and seen as unfit to govern by the secular urban elite. The grassroots conservative and liberal economic values that the party came to embody were automatically perceived as the antithesis of the secular, statist values upheld by the Kemalist establishment.[81]

Turkey's reform initiatives in the new millennium can thus be viewed within the framework of a socio-political power struggle between a secular urban elite, led by the Kemalist bureaucracy and the military, which has traditionally held power, and a rising middle class, represented by the AKP, which wants to play a larger role in shaping the country's future. Three factors gave the AKP the upper hand in this struggle during the party's first term in government, and enabled it to push through reforms with considerable success: a solid majority in the parliament, thanks to the strong electoral support of a large number of pro-EU Turks who saw in the AKP a promising alternative to the status quo parties; the backing of a group of liberal intellectuals who regarded the establishment's traditional control mechanisms over politics and society as the principle obstacles to Turkey's democratisation; and – fortunately for the AKP – a liberal-minded general at the helm of the TSK, General Hilmi Özkök, who, despite facing consistent criticism by hardline generals (and protests from junior officers), respected the authority of the elected government and allowed the reforms to take place.

Between late 2001 and early 2005, mostly in conjunction with the seventh and eighth harmonisation packages of the EU accession process, the AKP government carried out several reform initiatives that instantly made Turkey a more democratic country. These included the abolition of capital punishment and the expansion of minority rights and freedom of expression, as well as military reforms.[82]

In October 2001, the parliament amended the law governing the actions and status of the National Security Council, and removed the word 'priority' from the description of the council's recommendations to the government. The amendment further stipulated the appointment of more civilian members to the council, tipping the balance in favour of civilians. In August and December 2003, the parliament passed laws which set limits on the vast authority of the National Security Council general secretary and which made him answerable to the prime minister. The council was also ordered to meet every two months in future, rather than every month. In August 2004, a non-military man was appointed secretary-general of the council for the first time. A 2005 law ordered the withdrawal of the council's representatives from the Higher Council of Radio and Television, the Council of Higher Education, and the Turkish Radio Television Corporation.[83] Finally, the government is reported to have had more say than the military over the contents of the latest National Security Policy Document, drawn up in 2005.[84]

Significant advances were also made in reforming civil–military relations in the judicial and financial spheres. A 2003 law stated that civilians could no longer be tried in military courts. A law that took effect from January 2005 stipulated that there be parliamentary oversight of the military budget, and a February 2004 law paved the way for the Supreme Court of Accounts to 'supervise military expenditures and any extra budgetary resources', although this has yet to be ratified.[85]

In November 2004, the EU rewarded Ankara's reform initiatives by launching full membership negotiations. But observers within and outside Turkey remained split on whether Turkey's reforms would prove to be deep rooted or superficial. Those who argued the latter pointed to the military's continued interference in politics despite the constitutional adjustments. Larrabee and Lesser have called the 2001 changes 'largely cosmetic measures'.[86] İnsel argues that parliamentary oversight of the military budget did not lead to civilian authority over it, and that the National Security Council reforms merely made it possible to discuss more openly the military's position within the state, but did not in any effective way alter this position.[87]

When measured against the generic framework of a 'normal' democracy, in which the military is fully accountable to elected policymakers and focused exclusively on its normative duty of defending the country against external armed aggression, or – in a collective/regional approach – against the democratic standards set by the EU, Turkey's 'first generation' attempts at democratisation and military reform of the early 2000s clearly fall short of establishing substantive democracy. Nevertheless, by the same measures, these reforms stand out as breakthrough achievements; for sparking a broader genuine debate about the role of the military in Turkish society, and for being the first steps in dismantling the semi-authoritarian system put in place after the 1980 coup.

The picture since 2005: towards a 'New Turkey'

The bright picture of the early 2000s gave way to a bleaker one after 2005. The government undertook no significant reform initiative between November 2004, when membership talks with the EU began, and July 2007, when the AKP won a decisive victory at the snap parliamentary election it called following months of tense political stand-off over the presidential elections. The e-memorandum of April 2007, issued by the TSK at the height of the stand-off, was reminiscent of the old days, warning that secularism was under attack in Turkey.

Part of the reason for this downturn in the reform process has been the souring of relations between Turkey and the EU, fuelled by concerns and unfulfilled expectations on both sides. At the time of writing, negotiations remain partly suspended, and neither side is demonstrating the popular drive and the energy necessary to revive them.

By 2005, Turkey was once again facing a resurgence of PKK terrorism, which had largely died down by the late 1990s. This development was seen in Turkey as being connected with the growing security crisis in Iraq. As the media started again to broadcast footage of Turkish soldiers' funerals, which the public had been used to watching in the early 1990s, nationalist sentiment peaked across the country, and calls for securitisation took over from calls for further democratisation. By spring 2007, martial law had been declared in four provinces with a Kurdish majority, and troops were being massed along the border with Northern Iraq in preparation for a potential cross-border incursion to root out PKK bases there.

The populist tone that all parties adopted on the Kurdish issue turned increasingly adversarial as Turkey entered election season in 2007. The far-right Nationalist Movement Party and the main opposition Republican People's Party, which has the support of the military and the secular elite, repeatedly blamed the government for the revival of ethnic separatism and questioned the patriotism of the AKP leaders; the latter responded using similar nationalistic rhetoric.[88]

In the meantime, in August 2006, General Büyükanıt had succeeded General Özkök as the chief of the general staff. Özkök had been harshly criticised by hardline Kemalists for being too lenient on the Islamists in power and too accommodating of the EU reforms. Under his leadership, the TSK had acquired a significantly reform-minded image as an organisation that was respectful of civilian authority. Since his appointment,

General Büyükanıt, a hardliner, has frequently issued stark warnings about threats to secularism and the unity of the nation, thereby indirectly confronting the government.[89]

In March 2007, *Nokta*, a weekly political magazine, published a secret military document in which every prominent Turkish journalist was listed, along with their 'level of trustworthiness', as defined by their attitude towards the military.[90] A few weeks later, the same magazine published a diary that reportedly belonged to a retired high-ranked naval officer, which described in detail plans for two aborted coups against the AKP government made in 2004 by junior officers vehemently opposed to the curtailment of the military's influence.[91] In April, on the instruction of military prosecutors, the offices of the magazine were raided by the police. Eventually, *Nokta* was shut down and the editors were sued by state prosecutors for 'insulting and degrading the institutions of the state'.[92] Careful not to jeopardise its delicate relationship with the generals immediately before the presidential election, the government refrained from interfering in the episode, which exposed the TSK's continued meddling in civilian affairs, immunity from public criticism, and influence over the judiciary, and prompted Perihan Mağden, a journalist and an ardent critic of the military, to comment that 'even Hamlet's Denmark could not be so rotten'.[93]

The following month, in the midst of a tense political atmosphere, high in nationalistic fervour, the government attempted to replace Ahmet Necdet Sezer, the staunchly secular president, with Foreign Minister Abdullah Gül, whose background lies in political Islam. This met with the protests of the main opposition party, the military and the secular urban middle class, who feared that Turkey was drifting towards an Islamic regime. Massive pro-secular, pro-military rallies were organised in Ankara, Istanbul and Izmir, protesting not only against the government but also against the US and the EU. The military's e-memorandum signalled that the TSK was willing to block the democratic process in order to prevent an Islamist takeover of the presidency, which it saw as the last bastion of secularism. A crisis and a potential coup were averted when the AKP withdrew its candidate and called an early parliamentary election in July.

In the run-up to the July election, Turkey was frequently depicted both domestically and internationally as a country deeply and equally divided between the two ideological camps of secularism and Islamism.[94] Hence the clear victory that the AKP won on 22 July, securing nearly one out of every two votes, came as a surprise to many observers, not least the AKP leaders themselves. Emboldened by a renewed and strengthened popular mandate, the AKP confronted the military and the secular elite once again by naming Abdullah Gül as their presidential candidate for a second time, and, on 28 August 2007, Gül duly became the first president of Turkey to have roots in political Islam, as well as the first to be elected against the military's will.[95]

'The First President of the Second Republic', announced Ertuğrul Özkök, editor-in-chief of the daily *Hürriyet*, following Gül's election.[96] 'New Turkey' became a favourite phrase of political commentators, although there were significant variations in its connotations: the Kemalist press mourned the end of Atatürk's secular revolution and the looming establishment of an Islamic regime,[97] while the liberal media in Turkey, as well as much of the Western media, celebrated the triumph of democracy.[98]

It is certainly encouraging to see Turkey resolve a very sensitive – and potentially explosive – crisis through democratic means. There is reason to be optimistic about Turkey's democratic prospects, considering that the AKP, taking its performance up to late 2004 into account, has clearly been the boldest and the most successful reforming government in recent Turkish history. The 'New Turkey' could indeed be a more democratic one, if the new government makes good its pre-election promises to pursue EU-prompted reforms with renewed vigour, and to replace the 1982 constitution with one that allows for greater democratic rights and freedoms for all citizens, and ends the arbitrary authority of the state, without altering its secular character.[99]

One of the explanations put forward for the surprisingly decisive AKP victory is that the party won the votes of many secular urban Turks who objected to the undemocratic means employed by the military and the Republican People's Party to block Abdullah Gül's presidency. This idea prompted Hasan Cemal, a veteran journalist, to declare the election results 'the people's memorandum!'.[100] While there is a danger of sentimentally exaggerating this effect, its existence nevertheless suggests that there is increasing demand for substantive democracy in Turkey.

That being said, it is important to be wary of trying to draw definitive conclusions about the future of Turkish democratisation and military reforms from the July 2007 election results: the idea that Turkey has become fully democratic almost overnight – especially given the backdrop of the events preceding the elections – is just as far fetched as the suggestion that it has turned into an Islamic republic. Extreme nationalist sentiments, mixed with religious fanaticism and racism, have been rising across the country, evidenced by a growing number of attacks on ethnic and religious minorities.[101] As the 2007 *Transatlantic Trends* survey reveals, Turks are becoming increasingly xenophobic, overwhelmingly anti-American, and perhaps more crucially, against the EU, which has thus far been Turkey's main engine for reform.[102] At the same time, Turkey's Kurdish issue continues to seethe, as separatist attacks continue to take lives in Turkey's southeast, and the generals push for wider powers to deal with the situation. In October 2007, following a series of fatal attacks on Turkish soldiers and civilians, and in the face of mounting popular pressure, the Turkish parliament gave the military the green light to launch an operation into Northern Iraq.[103]

Above all, the Turkish military remains as immune to public criticism and as meddlesome in civilian politics as ever. On the eve of Abdullah Gül's election as president, General Büyükanıt spoke of 'sinister plans' and 'dens of evil' bent on breaking up the secular regime.[104] Immediately following the election, the generals adopted an explicitly confrontational attitude towards their new and democratically elected commander-in-chief.

The new era holds as many pitfalls as opportunities for the future of Turkish democratisation. Responsibility lies with the military and the secular opposition to respect and play according to the democratic rules; with the AKP to acknowledge that it no longer only represents its core Islamist constituency, but instead a much larger and heterogeneous coalition that demands service rather than ideology; and with President Gül to be a non-partisan head of state who can reach out beyond his devoted supporters to the whole nation, and especially to the highly apprehensive secular middle class. In what is essentially a political and economic power struggle between an urban elite and a broader, more mixed emergent middle class, there is hope – but no guarantee – that substantive democracy will prevail in Turkey.

CHAPTER THREE

Indonesia

Society

The military's perception of its role in Indonesian society: 'Pancasila'
and the dual function

When looking at any aspect of Indonesia, it is important to keep in mind the
territorial vastness and the cultural and ethnic diversity of this sprawling
archipelago. Made up of 17,508 islands covering a distance roughly equal to
that between London and Baghdad, Indonesia is home to over 230 million
people, four official religions, tens of ethnicities, and over 500 different
languages.[1] With its official motto 'unity in diversity', the country is a fasci-
nating case study for students of nationalism and nation-building.

Indonesia became a united and independent entity only after 1949,
following three centuries of Dutch colonial rule. Local armed militias
dispersed around the archipelago – mostly trained by the Japanese military during and after the Second World War to fight the Dutch, and united
under the leadership of a Javanese nationalist elite with the common cause
of overthrowing the colonial government – played a central role in the
founding of the Republic of Indonesia.[2] The Indonesian military, which was
formed from these armed militias, sees itself as the founder of the republic
and the hero of the nationalist revolution.[3] Its historical role is used by the
military to rationalise its traditional dominance in politics and society.

The Indonesian military (the acronym of its name in Indonesian is
TNI) has occupied a prominent place in Indonesian politics ever since

the republic's inception. It has directly intervened in politics twice: first in 1958, bringing to an end the short-lived parliamentary democratic era, and then in 1965, when the army, led by General Suharto, deposed President Sukarno, the founder of modern Indonesia, following a failed coup attempt on the part, allegedly, of the Indonesian Communist Party, and established the authoritarian 'New Order' regime.

In both 1958 and 1965, Indonesia was in the midst of political and economic turmoil and, according to some scholars, on the brink of dissolution were it not for the military.[4] In both cases, the military justified its intervention in terms of the objective of restoring 'order, stability and national unity'. This justification was generally thought to be validated by the military's role in founding Indonesia, and by its constitutional duty to uphold the official state ideology of *Pancasila*, or 'five principles'. Embedded as a set of clauses in the 1945 constitution, which was drafted during the nationalist struggle and is still in use today, these five principles encapsulate belief in one God (this was used against communism), national unity, social justice, 'civilised humanity' and 'democracy guided by consensus arising out of deliberations among representatives'.[5] After 1965, the interventions of that year and 1958 found their place in the official historiography as exemplary instances of the military's rescuing the nation from a variety of evils.[6]

The military reached the apex of its political and social influence during the New Order regime of President Suharto, having gradually established itself in a position where it was able to dominate all aspects of public life.[7] This domination was officially recognised and embedded in the doctrine of *dwifungsi*, or 'dual function', which allowed the military to act 'both as an agent of security and defense, and an important social and political force'.[8] The doctrine of *dwifungsi* was based on the image of a patriotic and altruistic military that had founded the republic and saved the nation several times from the treachery and incompetence of politicians.[9] It was an elitist ideology that betrayed a profound lack of trust in what the elite saw as the uneducated and ignorant masses, and their self-interested and narrow-minded political leaders.

During the New Order era, the military actively worked to enhance its second function of socio-political actor, under the pretext of safeguarding *Pancasila* and maintaining 'political order and social harmony', the two fundamental promises of the Suharto regime.[10] Its was an inflexible, narrow interpretation of *Pancasila* that left no room for criticism: put simply, in the military's view, you were either for *Pancasila* or you were against it, and, if against it, you were clearly an enemy of the state.[11] With the virtual annihilation of the Communist Party, and the military-sponsored massacre of hundreds of thousands of suspected communists throughout Indonesia between 1965 and 1966, communism did not pose a major challenge to the New Order regime for long. Instead, the military saw ethnic separatism, political Islam and liberalism as the imminent threats to its authority.[12]

According to Robert Lowry, an Australian army officer based in Indonesia in the 1990s, the Indonesian military elite regard ethnic separatism and political Islam as 'primordial sectarian forces'.[13] Both are deemed decentralising and destabilising elements, undermining the unity of the nation as well as the authority of the central government in Jakarta. The military has built a reputation for dealing brutally with separatist movements and repressing any political dissent in the periphery – especially in Aceh, East Timor and Irian Jaya (West Papua).

Regarding political Islam, the military elite opposes the idea of an Islamic state of Indonesia as being intrinsically against *Pancasila*, which, while including the belief in one God, also counsels toleration of religious and social plurality. The TNI also sees political Islam as a challenge to its socio-political authority. This perception has been particularly keen since the late 1980s, when Suharto began building strategic relationships with Islamist groups to counter the military's influence.[14]

Finally, liberalism, both political and economic, was seen as a threat by the New Order regime, as it had the potential to confront both the ideological monopoly of *Pancasila* and the *dwifungsi* role of the military. Lowry wrote in 1996 that 'liberalism poses the biggest threat to the unity of [the Indonesian military]. As education levels increase and the links between ideology, regime maintenance and personal power and privilege are increasingly exposed, ideological ardour weakens despite reinvigorated, community-wide indoctrination efforts.' He also noted that there had been repeated calls over a number of years from within the military for the organisation to remain united in the face of 'foreign and domestic forces intent on imposing *foreign concepts of democracy and human rights* and thus forcing [the military] back to the barracks'.[15]

On 21 May 1998, facing a deepening economic crisis, mounting international pressure and massive student protests, Suharto dramatically resigned, triggering a sudden and chaotic period of democratic transition known as '*reformasi*'; reformation. Only a few months later, the TNI announced that it would replace *dwifungsi* with a new doctrine; the 'New Paradigm'.[16] While the new doctrine proclaimed support for democratic reforms, and the military 'sharing' power with civilians and moving to the background in politics, it stopped short of accepting total civilian control over the security sector.[17] Although the TNI had seemingly abandoned *dwifungsi* overnight, it would prove more challenging for it to abandon its culture of influence.

Societal perceptions of the military: the engine behind reform

Three decades of direct involvement in government, and an appalling record of corruption, mismanagement and human-rights abuses, especially during the 1980s and the 1990s, gradually eroded the popular support and approval that the Indonesian military had enjoyed, to a degree, in the earlier decades of the republic. Thus increasing popular disillusionment with the upholders of the status quo was the main engine for reform in Indonesia at the turn of the millennium. In particular, a severe lack of popular trust in the military – demonstrated by massive protests throughout the country in the spring of 1998 – pushed the TNI towards instigating change. By the late 1980s, as Suharto's rule became increasingly repressive and corrupt, the military, seen in the country as the backbone of the regime, had gained a reputation as a brutal and unaccountable institution that hampered, rather than promoted, the nation's stability and development.[18] 'As an institution', wrote international economics scholar and Indonesia analyst Lex Reiffel in 2004, 'the TNI has been declining for twenty years or more'.[19]

In the country's periphery, the military came to be despised for its oppressive policies. In Aceh, which had enjoyed 500 years of independence prior to Dutch colonisation; East Timor, which the army had invaded for annexation to Indonesia in 1975; and Irian Jaya, whose population shares very little culturally and historically with the rest of Indonesia, the Indonesian military dominated virtually every layer of political and social life down to the village through its territorial command structure. Politics and development scholar Damien Kingsbury explains this as 'the means by which [the military] located itself throughout the country, placing units of soldiers at every level of society, paralleling each political tier of administration'.[20] This structure, which was, according to the military, the only way of keeping the country united, was perceived in the periphery as yet another occupation, similar to that of the Dutch, only more corrupt and brutal.[21] This resentment was further reinforced by the widespread conviction that the central government and the military were merely using the principles of *Pancasila* – the principle of national unity in particular – as an excuse to exploit the periphery's vast natural riches.[22]

The indiscriminate use of terror and force by the military when dealing with dissent carried anti-military sentiments to a nationwide audience.[23] The military typically put the blame on *oknum-oknum*,[24] or rogue members, for blatant human-rights abuses such as the shooting of hundreds of unarmed civilians in Dili, East Timor in 1991, and the countless atrocities reported in Aceh during the military's direct rule of the region between 1990 and 1998.[25] But such incidents ultimately cemented the military's deteriorating image as an undisciplined, unprofessional institution capable of inflicting harm on civilians.

Negative social perceptions of the military were also shaped by its extensive and frequently corrupt business interests. The TNI receives only a small portion of its funding from the central government, and most of its income comes from 'off-budget sources that no outside observer has been able to quantify'.[26] The military's financial interests range from big corporations at the national level to local businesses in the periphery. As Indonesia's economy started to liberalise from the 1980s onwards, it became apparent that most military-run businesses were uncompetitive and largely mismanaged. At the same time, stories of corrupt military officials and military involvement in illegal affairs became rampant.[27]

While the self-financing of the military has been socially accepted as a necessary practice in Indonesia since the first days of the republic, under the New Order regime, 'business' effectively became the military's third function.[28] Political scientist Harold Crouch has written that many officers felt much more at home dealing with Chinese-minority and foreign businessmen 'than commanding troops in the field'.[29] As a result, despite the military's projection of itself as altruistic and benevolent, and of civilian politicians as self-interested and corrupt, under the New Order regime, the TNI ultimately built a reputation for being single-mindedly focused on pursuing its own narrow interests.[30]

Above all, the military lost its credibility because of its role in propping up an increasingly authoritarian and unpopular regime. By the mid 1980s, the TNI had turned into a personal tool in the hands of Suharto, who successfully played rival factions within the TNI and the bureaucracy off against each other to consolidate power in his and his family's hands.[31] In the end, when Suharto fell – no longer able to deliver on his fundamental promises of 'stability and development' – the military, in the words of Asia analyst Adam Schwarz, 'found itself on the receiving end of widespread invective and hatred'.[32]

A growing conviction in the country that the regime no longer served the interests of the people ushered in a period of relatively free public criticism of the military and the government in the 1990s, known as *keterbukaan*, or 'political openness', which set the stage for post-Suharto reforms.[33] Crucially, these criticisms were also voiced within the military, which had been highly fragmented from the late 1980s onwards. During the 1990s, the Indonesian military became divided among officers loyal to Suharto, hardliners who criticised Suharto but supported *dwifungsi*, and reformers who criticised both Suharto and *dwifungsi*, and called for increased professionalisation of the military.[34] It was this last group who, encouraged by the popular desire for reform, democratisation and decentralisation (i.e., everything that the military had traditionally perceived as a threat), authored the New Paradigm and came to wield influence in national politics, as well as in the badly discredited military, in the late 1990s and into the new millennium.

Military and societal attitudes towards democratisation and military reform

For decades, Indonesia's military and bureaucratic elite interpreted 'democracy' however they saw fit, hollowing any concrete meaning out of the term. When Sukarno cemented his personal power by dissolving the elected parliament in 1958, he announced the beginning of a 'guided democracy'.[35] Similarly, in 1966, when the military established the New Order regime, Suharto named the new system '*Pancasila* democracy', an attempt to make the regime sound less authoritarian.[36] The military doctrine of *Tri Ubaya Cakti* ('Three Sacred Vows') of the 1960s stressed that the TNI did not seek unlimited power because as, in its own words, 'a freedom fighter', it 'always desires to be a constitutional force and a champion of democracy'.[37] These various formulations of 'democracy' tailored to personal and institutional needs were presented, in the words of Indonesian political scientist Pratikno, as 'democracy adapted to Indonesia's cultural context', while liberal democracy was labelled a foreign threat to political order and social harmony.[38]

Since the fall of Suharto, and particularly during the early 2000s, 'reform' and 'democratisation' have been popular catchwords in Indonesian politics. But what kind of democracy do Indonesia's civilian and military leaders espouse today, and what evidence is there that they are dedicated to *reformasi* as they claim? Until recently, after all, they vehemently opposed any liberal democratic changes of the kind that *reformasi* entails.

Critics of the TNI point to the steady decline in the pace of reform after an initial burst of reforming energy in 1998, the refusal of the military to accept total civilian control and its continued interference in civilian politics to suggest that the military tolerates democratic change only of a cautious and gradual kind, and only so long as it can control the pace and the parameters of this change.[39] When top military officers such as General

Ryamizard Ryacudu, the army chief of staff between 2002 and 2005, make such controversial comments as 'the meaning of civil supremacy is not that soldiers are under civilians, but that the whole nation, including the soldiers, has to obey the prevailing civil laws', or 'In Indonesia, we, the army, are part of the people. We cannot leave domestic issues to anyone', it raises doubts about the TNI's genuine commitment to untailored democratisation and military reform.[40]

Underlying General Ryacudu's remarks is a lack of confidence in the civilian population to manage the country's affairs that is arguably still shared by most officers in the TNI, and which helps the military fall short of espousing democratisation and military reform without conditions. Tata Mustasya of the *Jakarta Post* observes that most Indonesian generals have 'inflated ideas about their own leadership qualities'.[41] According to Indonesian political scientist Sukardi Rinakit, 'young officers are certain that Indonesians will need them sooner or later', and they are not happy to see their financial and political prospects fade.[42]

But while the military's definition of democratisation remains ambiguous at best, it is clear that there is a general desire inside the TNI to repair the organisation's badly tarnished reputation. Consequently, it is possible to claim that how earnestly the TNI attempts to reform will be largely determined by the public's continued enthusiasm and demand for democratic reform. At the turn of the new millennium, most Indonesians appeared to be heavily in favour of democratisation and military reform, but it was unclear what those terms precisely meant to the majority of the population.

Indonesia has made significant democratic strides since 1998: the military has officially abandoned *dwifungsi*; in 1999, free and fair elections took place for the first time in decades; people have enjoyed greater freedom of expression and information in the past decade than at any time previously; the legislative and executive branches have been significantly reformed and democratised.[43] Yet, the transition also proved to be a chaotic and uncertain time for most. As the traditional pillars of power of central government, the bureaucracy and the military, lost their tight grip on the political order, long-suppressed intercommunal conflicts resurfaced, and separatist movements gained momentum in the periphery.[44] At the same time, radical Islamist groups carried out attacks on Western targets in major cities – all this in a country still suffering from a devastating financial crisis.

When the initial euphoria of *reformasi* wore off, a lack of political and economic stability was what remained, which, according to political analyst Ikrar Bhakti, made many people nostalgic for Suharto's repressive but relatively orderly regime:

> What are the opinions of people on the streets? Their answers may surprise us. Many will say that they miss Suharto. During [the] Suharto era, according to them, security was the top priority, their daily income was higher than today and the price of daily necessities was quite low and certainly affordable for ordinary people. [45]

'Except for a small group of intellectuals, and pro-democracy activists, popular support for the current democratic system does not primarily reflect ideological commitment to democratic values as such', writes Dewi Fortuna Anwar, scholar and adviser to President B.J. Habibie, Suharto's successor:

> Instead, the majority base their enthusiasm for democracy on expectations that it will end the economic crisis, restore law and order, and provide them with a better opportunity to make a living. Given that people traded their political freedom for prosperity in the past, it is questionable whether they will now willingly trade their economic welfare for democracy. [46]

Adam Schwarz suggests that, by the 1990s, the memory of the 1950s, and Indonesia's only experience of Western-style democracy, had mostly faded, and all that remained of the democratic era was a memory of bitter socio-political divisions, kept alive for over three decades 'by tireless reminders [from] Suharto's government'. [47]

The modesty of Indonesians' desire for democracy may well also be partly due to doubts about the effects of recent reforming measures. Decentralisation laws brought in in 1999, for example, have proved a disappointment to many. Pratikno notes that, while the laws achieved a significant transfer of power from the central government to the formerly feeble provincial administrations, they did not necessarily introduce good governance: 'while corruption in Jakarta has not been reduced, the spread of corruption to the local level has increased significantly'. [48] Sociologist Vedi Hadiz quotes a provincial politician in Central Java arguing that 'because the culture of bureaucracy remains the same, the decentralization of power and authority will be followed by the decentralization of [corrupt] practices'. [49]

Yet these observations should not lead one to conclude that substantive democracy is a lost cause in Indonesia. Democratisation is a long and difficult journey in uncharted waters for a politically and economically fragile country like Indonesia, which has been under a dictatorship for over three decades. It is unrealistic, even unfair, to expect democratic reforms to have taken root in society in less than a decade. The evolution of a culture of democracy is a much more gradual process, which, despite a great number of obstacles, Indonesia has the potential to achieve.

The global order

Global influences on Indonesia's military reforms and democratisation
When Indonesia's military leaders established the New Order regime, they promised political order and economic development, in contrast to Sukarno's goals of 'continuing the 1945 revolution' and 'struggle against imperialism'.[50] In line with this, the country's new leaders abandoned Sukarno's policy of *konfrontasi* ('confrontation') with its neighbours, Malaysia in particular, over ideological and territorial disputes.

Fuelled by a conviction, widely held, that the greatest threats to Indonesian national security came from within the country – in the form of separatist movements, religious radicalism and other challenges to *Pancasila* – the military elite acknowledged that they had no time or energy for a confrontational foreign policy.[51] Throughout the lifespan of the New Order regime, and arguably well into the democratic era, the primary focus of foreign policy has been attracting foreign capital into the country, in the shape of aid and investment, and finding markets for Indonesian exports.[52]

The Cold War saw the division of Asia between communist and liberal economies. While this ideological divide ran deep in Indonesian politics and society during Sukarno's rule, its relevance ended with the closure of the Communist Party and the bloody anti-communist crackdown of 1965–66.[53] Despite Suharto's increasingly autocratic rule, Indonesia came to be regarded as a relatively harmonious and peaceful country internationally in the New Order years, at a safe distance from the main fronts of the ideological battle being waged between the US and the Soviet Union.

Indonesia's rather peaceable foreign affairs constitute a sharp contrast with its tumultuous internal affairs. It was these internal tensions, rather than any outside influences, that brought the greatest pressure to bear on the authoritarian regime and the military, pressure which ultimately caused the first to collapse and the latter to accept reform. It is important, however, to identify one external trend that has supplemented the internal push for democratic reform: globalisation.

Following the collapse of the Soviet Union, democracy and human rights became central preoccupations for the global community.[54] The Indonesian military's continued disregard for human rights and good governance, and support for a regime that failed to couple economic liberalism with political liberties, put it at odds with the changing global values. 'To cope with such changes', writes Sukardi Rinakit, 'the military needed to redefine its socio-political role. Otherwise, Indonesia would be shut out of the global community and the international organizations would not support it if the country was faced with problems.'[55]

It would be wrong, however, to assume that the TNI adopted the New Paradigm simply to impress the international community. The globalisation of democratic values also made its mark on Indonesian society, especially among the better educated younger generations, who had not experienced the historical events that legitimised *dwifungsi*. This put the military under substantial social pressure during the *keterbukaan* era, pressure which reached a peak of intensity at the end of the millennium.[56]

The economic effects of globalisation also propelled the push for reform in Indonesia. As we have seen, the country's participation in free-market competition put the spotlight on the military's corrupt and mismanaged business activities. Rising expectations of business efficiency in society have meant that military leaders have been increasingly urged to make use of the technical and managerial skills of civilian professionals in their financial dealings.[57]

Finally, the Asian financial crisis of 1997 played an instrumental role in accelerating democratisation in Indonesia. It has often been speculated that, were the Indonesian economy to have continued to grow at 6–7% per annum, the growth rate before the crisis, after 1997, authoritarian rule would have been superseded by a more democratic system in 15 or 20 years.[58] As it was, change came more swiftly. The drastic fall in the value of the rupiah resulted in a steep rise in food prices, leading to massive riots and a great loss of confidence in the regime.[59] According to Dewi Fortuna Anwar, the crisis 'struck at the very heart of the New Order's *raison d'être*' of economic development: it brought out in the open the 'simmering resentment' that existed over the widening economic gap in society, and 'united and emboldened' the middle class and a nascent civil society, as well as making Indonesia more dependent on external assistance, and hence more vulnerable to international pressure to democratise.[60]

Regional influences

While globalisation has been influential in Indonesia's democratisation, it is interesting to note that Jakarta has had no deep commitment to any foreign country or international organisation that has urgently necessitated undertaking democratic reforms, in the way that Ankara, for instance, has had with the European Union. The lack of an external 'democratic anchor' is especially evident in the country's regional relations.

Indonesia is one of five founding members of ASEAN, a geopolitical and economic organisation established in 1967 as a front against communism. Southeast Asia analyst Joakim Ojendal has credited ASEAN with being 'perhaps the most advanced and successful case of regionalization in the South'.[61] Dewi Fortuna Anwar has said that ASEAN's success in providing stability and harmonious relations in the region is 'in stark contrast to the earlier period before the association was founded'.[62]

ASEAN now encompasses all the nations of Southeast Asia; in the 1990s it expanded to include Vietnam, Myanmar, Laos and finally Cambodia in 1999.[63] In addition to its own regular summits, the organisation also provides a platform from which member states can engage with the wider world, through the ASEAN Regional Forum, ASEAN+3 and other forums.[64]

However, ASEAN's long list of achievements, which includes the promotion of regional security and stability and economic cooperation, and the provision of greater international recognition and a stronger bargaining position to member states, does not include the promotion of democratisation and military reform. In fact, the organisation has often been criticised for tolerating undemocratic regimes, such as the junta-led Myanmar and Thailand. On the issues of democracy and civilian control of the military, ASEAN is considered by many to be a mere 'talk[ing] shop'.[65] It is telling that, at present, Indonesia is arguably ASEAN's most democratic member state.

Vedi Hadiz has remarked that Indonesia's neighbourhood lacks solid models of democratisation. 'In Indonesia, Thailand, and the Philippines', he writes, 'social forces that may have an interest in breaking up the stronghold of predatory elites are marginalized in the democratic arena because they are not equipped to play the games of money politics and political intimidation.'[66] According to Hadiz, only South Korea – with all the faults of its system – can provide an example for Indonesia to follow. But for Indonesian policymakers looking for a success story in the region to emulate, the example of a regime such as this is liable to be overshadowed by more exciting examples of spectacular economic transformation, such as China, which rarely display accompanying democratic progress.

Western influences

The role of Western liberal democracies in Indonesia's democratisation process has been mixed. The potentially normative influence of the European Union in Southeast Asia has frequently been negated by its economic interests. The recent decision by the EU to launch free-trade negotiations with ASEAN, despite the obvious problems of democratic governance in some of its member states, is one example of this.[67] A particularly striking indication of the EU's priorities in the region came in 1999, when Brussels lifted a ban on arms sales to Indonesia only four months after the TNI-sponsored virtual destruction of East Timor followed a referendum calling for independence.[68] In any case, the EU is too far away from Indonesia and has never been concerned enough with its internal affairs to have a strong normative influence on any reform process.

Surprisingly, although it is considerably nearer to Indonesia than Europe is, Australia can be said to show a similar lack of concern. Gareth Evans, the Australian foreign minister from 1988 to 1996, admits that Australia's past cooperation with Indonesia's military 'helped only to produce more human rights abuses'.[69] Bilateral relations have become increasingly sour over the past decade. With issues such as territorial disputes, the terrorist attacks on Australian tourists in Bali and the fate of Papuan refugees in Australia filling the agenda, the relationship hardly has the feel of a cooperative relationship between fellow democracies.[70] This deterioration further diminishes any influence Australia might have.

It is the United States that has generally had the most direct impact on Indonesian politics. However, Washington's influence on Indonesian democratisation and military reform has been inconsistent and controversial at best. The staunchly anti-communist New Order regime made Indonesia, in the words of scholars Angel Rabasa and John Haseman, into 'a pillar of the US-backed informal regional security system'.[71] In return, throughout the Cold War, the US was the primary provider of military assistance to Indonesia, tacitly enabling the regime's occasional brutality against its own citizens.[72]

As Indonesia's strategic importance disappeared with the end of the Cold War, Washington could no longer disregard Jakarta's human-rights abuses and authoritarian rule. In 1992, US Congress cancelled $2.3bn in military assistance to Indonesia, following the killing of over 200 civilians by the TNI in East Timor a year earlier.[73] In 1994, Congress imposed a weapons embargo, which stayed in place for over a decade. During the Asian financial crisis, the US played a significant role in ousting Suharto by tightening the arms embargo and making financial aid (both direct and via the IMF) conditional on the instigation of urgent democratic reforms and an end to military interference in politics.[74]

However, Washington's stance towards Indonesia changed once again after the 11 September attacks. The Bush administration views Indonesia as a key ally in its 'war on terror', and has pushed Congress to ease restrictions on military assistance.[75] In 2002, Paul Wolfowitz criticised the Indonesian government for not making enough progress on fighting terrorism, while praising Pakistan's authoritarian military ruler General Pervez Musharraf for his cooperation on this front.[76]

In 2005, in response to increased support from Jakarta, particularly in clamping down on radical Islamist cells inside Indonesia, Washington announced that it would lift its arms embargo.[77] Some analysts saw this decision as a recognition of Indonesia's security-sector reforms.[78] Others, including human-rights activists, NGOs and pro-democracy scholars, considered it a major setback for justice, human rights and further reform, and a green light to the TNI to pursue its political and economic interests across the archipelago with impunity, while hiding behind the cloak of national security.[79] 'Most observers in Jakarta', writes Bhakti, 'believe that the present Bush administration is more focused on the war on terrorism than on democracy.'[80]

The US has probably had more leverage than any other external actor over Jakarta's policy decisions. It is difficult to argue, however, that Washington has consistently used this power to push for democracy in Indonesia. Recent history demonstrates that Washington only becomes a genuine supporter of Indonesian democratisation and military reform when these goals do not clash with (or in fact serve) its various strategic interests in the region.

State

Undoing the New Order legacy: achievements and shortcomings of military reform

Successive Indonesian governments have attempted to dismantle the authoritarian system put in place by the military during the New Order era. Some reforms have been effective, while others have appeared rather half-hearted. Outlined below are the most important initiatives geared towards reforming the TNI, and their apparent results.

Political reforms

Under the New Order regime, the TNI exercised profound influence in both the legislative and executive bodies, as well as the judiciary, which was achieved by subordinating it to the executive branch.[81] High-ranking officers were appointed to and dominated the People's Consultative Assembly (MPR), the country's highest representative body, whose function is to uphold and amend the constitution, inaugurate the president and determine the national agenda.[82] The MPR has three principal components: the People's Representative Council (DPR), an elected body, the Regional Representative Council (DPD), and a chamber of 'functional representatives' of important social groups, professions and institutions.

Before the 1999 elections, there was a substantial contingent of military officers (and their wives) in the DPD, and the governors of a number of provinces were army generals.[83] Until 1997, besides holding permanent seats as a 'functional group' in the MPR, the military also held 100 out of 550 seats in the 'elected' DPR, an arrangement which, according to Dewi Fortuna Anwar, was viewed as 'compensation for not being allowed to vote'.[84] In all, unelected military officers routinely occupied at least a third of the legislature and the executive, exercising direct control over Indonesia's political affairs.

The New Paradigm obliged the military to take a backseat in politics. As a part of this withdrawal, the TNI announced its disassociation from Golkar, Suharto's ruling party, in 1998.[85] In the electoral campaign that quickly followed the laws coming into force in February 1999 that provided for a 'free and fair' election, the military's control of the legislative and executive branches became one of the top campaign issues.[86] After the election, a reform-minded and democratically elected government reduced the number of seats allotted to the military in the DPR to 38.[87] A 2004 law made DPR membership conditional on resignation from any official posts and election. In the meantime, the percentage of the DPD made up of military officers was reduced from 20 to 10.[88] As a result of these reforms, and despite the fact that the TNI still has permanent functional representatives in the MPR, the military's direct control over Indonesian politics has waned significantly.

Economic reform

The Indonesian military has had a funding problem ever since the founding of the republic. Defence spending in an average year amounts to less than 1% of Indonesia's GDP,[89] covering under a third of the armed forces' operational costs.[90] The business deals that the military has made in order to cover the remainder of its expenses have caused the country significant problems. After many years of corruption, mismanagement and lack of transparency under the New Order, the military's involvement in business has become a major burden on the national economy.[91] It has also deeply tarnished the organisation's reputation, and reduced its overall efficiency and professionalism.[92]

According to Sri Yunanto, director of the RIDEP institute in Jakarta, which focuses on security-sector reform, there are three kinds of military business arrangements at work in Indonesia: 'institutionalised' (managed under foundations or corporations); 'non-institutionalised' (run individually by retired officers); and 'grey businesses' (in which military personnel provide protective services for illegal activities).[93]

Scholar and political activist Ichsan Malik notes that 'corruption, nepotism and exploitation' are still rampant in the democratic transition period.[94] In April 2006, Indonesian magazine *Tempo* reported that top army officials had diverted some $2.4m in government funds in 2003.[95] According to Danang Widoyoko of Indonesia Corruption Watch, it is well known that the military is involved in the marijuana trade in Aceh, where it is also believed to accept donations from the oil company Exxon Mobil for 'business protection'.[96] In East Kalimantan, the military involvement in illegal timber business is, according to one Indonesian newspaper, 'so well-known that a saying has emerged that the soldiers in the region carry not rifles but electric chainsaws'.[97]

Before 2004, successive Indonesian governments, under pressure from the TNI, had shied away from attempting to reorganise military business arrangements and tackle military corruption. But in 2004, soon after assuming office, the government of President Susilo Bambang Yudhoyono passed a law banning military business activities and requiring the government to take over all military businesses by 2009. While this might seem to be a law of historic significance, critics have pointed to numerous exceptions hidden in the footnotes that will allow many businesses to remain with the military.[98] Another concern is the slow pace of implementation: three years after the enactment of the law, no company has yet been taken over by the government.[99]

Internal security reforms

During the New Order era, there was no distinction between the military's external and internal duties, and the police were a part of the armed forces, often taking a backseat to the army in security affairs.[100] According to Dewi Fortuna Anwar, this led to an 'excessive security approach' to law and order.[101]

In 1999, the post-Suharto government formally separated the police force from the armed forces, and made the police an independent institution directly answerable to the president.[102] Although the police lack the personnel and funding necessary to be fully independent and efficient in internal security matters, and thus often seek out the army's assistance, the separation was nevertheless an important move towards narrowing the TNI's focus to its normative duty of defending the country from foreign aggression.[103]

An area in which there has, controversially, been much less movement is in weakening the structure of the military's power across Indonesia's regions. The TNI's territorial command structure, whereby appointed military officers parallel elected civilian administrators on the regional level across the archipelago, was envisioned as a way of controlling the internal security situation and upholding national unity. It ensured the military's – unwelcome – domination over the socio-political affairs of the provinces throughout the New Order era.[104]

Critically, the territorial command structure remains firmly in place, and its maintenance was further consolidated in the 2005 defence white paper in the name of counter-terrorism.[105] Many officers strongly oppose disbanding the structure: there is a persisting belief within the military that it is the only way to keep the country united and the TNI relevant.[106] Despite repeated claims by the government that the military will not use its territorial command structure to regain power nationally, critics argue that without removing or substantially reforming the structure, a total abandonment of *dwifungsi* cannot be possible.[107]

The president vs the military: the delicate balance of civil–military power

A good way of monitoring the pace and direction of Indonesia's military reforms, as well as its civil–military balance, is to observe the fluctuating relationship between the presidency and the military. Suharto's divide-and-conquer strategy for keeping the military under his personal control created a culture of factionalism and endemic rivalry between the presidency and the military, as well as within the ranks of the military. While the political influence of both institutions have been weakened relative to the New Order era in the wake of *reformasi*, this culture of factionalism and rivalry between them has survived.

Indonesia has had four presidents since Suharto. The first, B.J. Habibie, of Suharto's Golkar party, was appointed as interim president after his fall. Despite not having been elected, and being a former vice-president to Suharto, Habibie undertook significant political reforms within a year of coming to power, including passing the Political Parties Law, which allowed parties to run freely for election, the Regional Autonomy Law enabling political and economic decentralisation, and a law liberating the press.[108] However, his efforts to release his former boss, Suharto, from corruption charges marred Habibie's otherwise good track record.[109]

In 1999, Habibie surprised many Indonesians, not least his supporters in the TNI, by announcing that a referendum to choose between autonomy and independence would be held in East Timor. This decision, coupled with his abolition of direct military rule (the 'military operation zone') in Aceh, made him extremely unpopular with the military, who thereafter withdrew all support for him.[110] When the people of East Timor voted overwhelmingly for independence, pro-Indonesia militias, armed and sponsored by the TNI, unleashed the campaign of violence that almost annilated East Timor's entire infrastructure, and forced the population to flee to the mountains.[111]

The same year, Habibie was succeeded by Abdurrahman Wahid, also known as Gus Dur, in the first free and fair election in Indonesia's history.[112] Wahid, who also initially enjoyed the support of the TNI, seemed determined to continue with the military reforms of *reformasi*. To this end, he allied with General Wiradikusumah, an outspoken reformist officer who led the 'radical' faction of the TNI.[113] The president was applauded by pro-democracy groups when he fired a powerful former supporter of his, the coordinating minister of politics and security, General Wiranto.[114] A deeply influential officer and the leader of the conservative faction of the TNI, Wiranto was widely held responsible for orchestrating the human-rights abuses in East Timor. In 2000, the Commission of Inquiry into Human Rights Violations in East Timor, established under Gus Dur's presidency to bring perpetrators of past human-rights crimes to justice, found Wiranto and five other generals culpable 'for a wave of killing and destructions' in 1999.[115] However, the trials ultimately exposed the military's enduring influence on the judiciary, as Wiranto was later acquitted, and ran for presidency in 2004.[116]

Wahid's relationship with the TNI was further strained in 2000, when the military's role in fuelling intercommunal violence in the Moluccan Islands was uncovered.[117] In the meantime, the president became embroiled in two financial scandals that damaged his reputation, and gave his opponents in the TNI a bargaining chip. The scandal enabled the military first to force him to fire General Wiradikusumah, who had been marginalised within the military for his role in unveiling corruption in the TNI involving

$20m.[118] Then, in 2001, following a joint campaign by his civilian and military opponents to publicly discredit the president, and with tanks surrounding the presidential palace, Wahid was unanimously impeached by the MPR. According to Damien Kingsbury, 'the move to unseat the president was underscored by interests that were less concerned with "democracy", much less "reform", and more to do with re-establishing the power of long-standing vested interest.'[119]

Gus Dur's deteriorating relationship with the TNI, and his eventual demise at its hands, sent a strong signal to his appointed successor, Megawati Sukarnoputri, the daughter of Indonesia's first president, Sukarno. During her term, reform initiatives came to a virtual standstill. The 2005 report of the Bonn International Centre for Conversion on Indonesian security-sector reform noted that:

> Megawati appears to have adopted a 'no policy' stance, especially in the area of reform of Indonesia's security apparatus, in order not to alienate the military in her approach towards internal conflicts and against terrorism. In addition, her inability to tackle further the continued problem of widespread corruption eventually led to her political demise.[120]

In 2004, Susilo Bambang Yudhoyono, a retired general and a reformist, was democratically elected to replace Megawati, promising reinvigorated reform and democratisation. Some observers believed that his status as a former senior officer and his track record of fighting corruption would enable him to bring the armed forces fully under civilian control.[121] Others remained sceptical: 'To be realistic, Yudhoyono was nurtured under the repressive New Order regime, and like Soeharto was an Army general', wrote Harry Bhaskara of the *Jakarta Post* in 2006. 'A horse cannot breed a dove, as they say.'[122]

Yudhoyono has had some significant achievements, the most important of which was the signing in August 2005 of a peace treaty extending autonomous rights to the provincial government of Aceh, which put an effective end to the three-decade-old conflict between the central government in Jakarta and the separatist Free Aceh Movement.[123] He appointed to the position of defence minister Juwono Sudarsono, a prominent critic of the military's business dealings, who subsequently championed the 2004 law that mandated an end to military involvement in business.[124] Yudhoyono also appointed a reformist officer, Djoko Suyanto (unconventionally, from the air force), to lead the TNI. Under Suyanto, the TNI issued a new doctrine, *Tri Darma Eka Putra*, in 2007, which emphasised the military's disassociation from politics.[125]

But the reforms under Yudhoyono have been characterised by many as half-hearted and too slow – a common criticism of 'second generation reforms' in democratising countries. As Rabasa and Haseman concluded in 2002, the TNI 'still exercises political influence at national and regional levels and has the capacity, although currently not the intent, to recapture the political heights'.[126] As we have seen, the military's involvement in corrupt and illegal businesses is still very much a reality, and the controversial territorial command structure remains firmly in place. For Indonesia analyst Jun Honna, the military reforms simply mean that old wine is served in new bottles: '[the TNI] has adapted to the new, emerging political landscape and honed its political skills in accommodating, deflecting and manipulating pressures for change in ways that have enabled it to maintain political power'.[127]

Concerns about the outcomes of Indonesia's democratic reforms are not limited to those that deal with the military. As we know, political decentralisation has not put an end to institutionalised corruption. It is also doubtful whether decentralisation has meant more rights and liberties for the population at large. According to Syafi'i Anwar of Indonesian think tank the International Centre for Islam and Pluralism, by mid 2005, 18 regional governments had adopted highly restrictive sharia laws, and 13 others were in line to do so.[128] The rise of religious conservatism in the provinces has been coupled with the rise of political Islam at the centre. Freed by the political liberties granted under *reformasi* from the tight control of the *ancien régime*, hardline Islamist groups have been using democratic platforms to push for nationwide limits on individual liberties, such as the establishment of a 'moral police', restrictions on clothing, the criminalisation of 'unethical' behaviour such as kissing in public, and the insertion of a clause about sharia into the constitution to reverse the secular principle espoused in *Pancasila*.[129]

Justifiable though concerns may be, however, the progress Indonesia has made over the past decade is nevertheless commendable. Almost overnight, an authoritarian country with liberal tendencies turned into a democratic country with authoritarian habits. The country's progress is especially impressive considering that, from a collective/regional perspective, it has virtually no external anchor against which its progress may be measured. When judged against the normative ideal of achieving substantive democratisation, it is clear that much work remains to be done, as democracy is still in its nascent stages and remains fragile in the face of challenges such as ethnic separatism, religious fanaticism, economic crises, widespread corruption and a military that is highly meddlesome in civilian affairs. But there is reason to be optimistic, so long as future Indonesian governments stay on the track of reform and remind the military that its actions will not go unchecked as before, and the people press both institutions to keep the spirit of *reformasi* alive.

CHAPTER FOUR

Comparative Analysis and Conclusions

A comparative analysis of Turkish and Indonesian experiences of military reform and democratisation

As we have seen, despite being half a world apart from one another, Turkey and Indonesia share certain basic characteristics: both are non-Arab nations with predominantly Muslim populations, which have influential militaries that have been at the centre stage of social and political life since the birth of the republic. Moreover, both Turkey and Indonesia embarked with unprecedented energy upon a process of democratisation and military reform at the turn of the millennium. Looking at the two countries in 2007, it is clear that the energy driving reforms in the late 1990s and the early 2000s has been in gradual decline. Nevertheless, while their militaries remain highly influential in civilian politics, it is fair to say that both countries are more democratic today than they were a decade ago.

What can we make of Turkey's and Indonesia's experiences of military reform and democratisation? Let us draw together what we know of each to see what lessons may be derived from their collective experiences.

The socio-political role of the military

A fundamental characteristic of both the Turkish and the Indonesian militaries is their perception of themselves as virtually sacrosanct. This perception underpins the traditionally central socio-political position of the military in both countries. Officers in Turkey and Indonesia see themselves as the heirs to the founders of the republic, and as the main architects of their respective countries' nation-building projects; a view that gives both institutions a sense of ownership over the state and of a right to direct the country's political affairs. This sense of ownership is embedded in the official ideologies of Turkey and Indonesia upheld by the military, Kemalism and *Pancasila*, which confer on the military the duty of guiding or supervising the nation and protecting the regime against all threats, internal and external.

Two complementary beliefs, shared by both the TSK and the TNI and bolstered by each country's bureaucratic elite, underlie the sense of a military duty to supervise the people: a self-confidence which presupposes that the military knows what is best for the people, because it is the only institution that is truly patriotic and altruistic; and a severe lack of confidence, or mistrust, both in the general population, based on the assumption that they are ignorant and misguided, and in politicians, because they are seen to be self-interested and easily corruptible. Consequently, both the TSK and the TNI feel entitled to routinely intervene in civilian affairs 'in the interest of the people'.

In placing themselves at the centre of their country's socio-political structure, the TSK and the TNI encourage the militarisation of society and politics. Military officers in both countries are trained to consider any idea that originates outside the military's control and challenges the official ideology as a potential threat to the regime. This outlook makes the TSK and the TNI intrinsically suspicious of and resistant to ideas of military reform and democratisation.

Societal perceptions of the military

Perhaps the greatest contrast between these two case studies is between the social perceptions of the military in each country. Whereas in Turkey, the majority of the population views the TSK as a disinterested, efficient and incorruptible institution, in Indonesia the popular view is quite the opposite. While for Turks, the military is the most trustworthy national institution, Indonesians rank theirs among the least trustworthy.

History and cultural differences offer part of the explanation for this contrast. Turkey is the successor state of a larger independent entity, the Ottoman Empire, and its army – the empire's most established and highly valued institution – has traditions that date back centuries. In Indonesia's case, never before Dutch rule did a state exist which controlled the vast territories that are Indonesia today. The institution of the Indonesian military, like the Republic of Indonesia itself, is a product of the twentieth century. Many of those provinces of Indonesia that had previously been independent therefore perceive the central government and the military as new occupiers; a view which challenges the military's self-perceived legitimacy.

Another reason for the contrasting views can be seen in the difference between the TSK's comparative efficiency and apparent incorruptibility and the TNI's corruption and lack of discipline. Here, one should remember the difference between the two organisations' economic conditions, and in particular the minimal state funding received by the TNI, in spite of the monumental cost of financing the security of the vast Indonesian archipelago. While military officers are the best paid and trained state servants in Turkey, Indonesian soldiers are seriously underpaid and are encouraged to find their own means of income. As the better paid and better trained military, it is only natural that the TSK is able to project a more disciplined and altruistic image than the TNI.

Finally, the Turkish military has largely refrained from having any long-term direct involvement in politics, and has been careful not to identify itself with any party or leader. It has returned power to civilian administrations after almost all of its interventions, and has instead chosen to influence society and politics relatively discreetly through a variety of institutions, such as the National Security Council – which is typically portrayed as being nothing more powerful than a consultative body – and organisations such as the Higher Council of Radio and Television and the Council of Higher Education. This has not been the case with the TNI, which has been directly involved in government for over three decades, and has been affiliated with a political party (Golkar) and its leader (Suharto).

Popular attitudes towards military reform and democratisation

While disillusionment with the TNI in Indonesia, which brewed under the New Order and reached its zenith during *reformasi,* made it the focus of demands for reform at the turn of the century, most Turks remain bemused as to why their country's most trusted and successful institution needs to change.

In Indonesia, opposition to the TNI's direct involvement in politics and business gradually opened the way for the people to criticise it freely, whereas in Turkey, publicly criticising the TSK is still considered unpatriotic, and is more or less taboo. Today, while Indonesian society has become the main engine of that country's drive for military reform, most Turks are either indifferent or hostile to military reforms, which they typically regard as unrelated to the issue of democratisation.

A deeper question persists as to what the majority of Turks and Indonesians understand by democratisation. There is reason to believe that, when most Turks and Indonesians express support for democratic reforms, they are really communicating a desire for political stability and economic development. It should be noted that, in both countries, popular demand for democratisation peaked in the wake of devastating economic crises that were associated with widespread corruption and mismanagement. As the economy stabilised, the momentum for reform in each country gradually declined.

In both Turkey and Indonesia, a relatively small – but growing – proportion of the population does appear to be demanding proper or substantive democratisation. In Turkey, these are the people who protested undemocratic interference in the political system in the run-up to the presidential election. In Indonesia, they are the activists, scholars and civilian leaders who strove to keep democratisation on the agenda when the drive for securitisation and traditional concerns over stability seemed to be displacing it. But despite the existence of this constituency, the tailored interpretation of democratisation by the majority of the population in both countries ultimately helps reforms to remain shallow and makes their consolidation more difficult.

Principal drivers of reform

Turkey lacks the popular enthusiasm that can be seen in Indonesia for reforming the military, yet Indonesia lacks an external anchor, such as the EU has been for Turkey, to push it further down the track of democratisation. Indeed it can be argued that Indonesia has no immediate role model, let alone a formal commitment to a regional or international organisation, that might encourage its government to instigate deeper reforms. Regionally, ASEAN has failed to play such a part, as has Australia, despite its geographic proximity.

On the other hand, Indonesia's relative geopolitical isolation may turn out to be an advantage from the point of view of reform. Located at the crossroads of three continents, and surrounded by divergent cultures and ideologies, most Turks, and the military and the Kemalist elite in particular, seem to have grown highly suspicious of foreign interests in Turkey, including those pushing for democratisation; a resistance that is not so much in evidence in Indonesia.

The Turkish case also illustrates that unless there is a widespread internal desire for reform – that is not limited to economic development – the influence of an external anchor alone will be inadequate. Furthermore, the growing lack of popular support for Turkish admission to the EU within Europe itself suggests that relying on the unwavering support of an external agency to carry out internal reforms may not be a wise strategy on the part of a country's reformers.

The effects of globalisation and the post-Cold War order

An external factor that has had profound social influences on both countries and helped trigger reform is globalisation. The end of the Cold War seriously challenged both the TSK and the TNI. The increasing international emphasis on liberal values, democracy and human rights, coupled with calls for institutional openness, transparency and accountability as a result of growing economic integration into competitive international markets, confronted the static, state-centric and elitist ideologies of the Turkish and Indonesian militaries, as well as the undisputed sovereignty of the nation-state. At the same time, as populations became more affluent, the number of pro-democracy activists, scholars and intellectuals grew.

Indonesia responded to the globalisation of economies and liberal values by ending Suharto's autocratic rule. The TNI, in the face of pressure, claimed to have abandoned the doctrine of *dwifungsi*. Amid shifting global values and alliances, motivated by the hope of finding the economic and political stability it had never previously enjoyed, Turkey sought a reconfirmation of its place in the Western world by making full membership to the EU its primary national aspiration. The same pressures seem to have had a mixed effect on the TSK, as the organisation is at once significantly cooperative with global institutions, especially the UN and NATO, and highly resistant to the new global order.

Globalisation has not always encouraged democratic reform. The globalisation of Islamist terrorism and the 'war on terror' following the 11 September attacks have worked against the democratic current in both countries. They have led governments to prioritise securitisation over democratisation, and introduced a sense of urgency that has overridden calls to establish complete civilian authority over the military.

The role of the United States

The United States has historically played a controversial role in both Turkish and Indonesian democratisation and military reform processes. Cold War rivalry and a concern for geopolitical stability prompted Washington to lend its support to Indonesia's authoritarian New Order regime, and to back Turkish military coups, thereby delaying both countries' democratisation.

But with the end of the Cold War, as the US focus shifted from Indonesia's strategic importance to Suharto's autocratic rule and the TNI's blatant human-rights abuses, an arms embargo and aid conditionality accompanied the mounting internal pressure on the regime to eventually bring about the fall of Suharto. In Turkey, while Washington has routinely expressed its support for Turkey's democratisation, and backed its EU bid, it has been careful to refrain from criticising the military's interventions in civilian affairs, in order to avoid alienating a powerful regional ally of its own military.

Under the current Bush administration, effective American support for democratisation and military reform in both Turkey and Indonesia appears to have been largely replaced by a new type of strategic approach, reminiscent of the ideological alliances of the Cold War, which prioritises securitisation and regional stability over democratic reform. A perception in both countries that America has failed to address the issue of Palestine, coupled with opposition to its invasion of Iraq with the purported goal of democratisation (and, in the case of Turkey, a feeling that the US has failed to adequately support Turkey's own 'war' against terrorism) have together made not only for an unprecedented increase in anti-American sentiment in both societies, but also, crucially, to a loss of faith in the urgency of democratisation.

Conclusions: observations from Turkey and Indonesia of military reform and democratisation

External support helps to enable reform, but internal drive is crucial
As both cases suggest, external commitments and pressure, carrots (e.g. conditional financial aid, or the prospect of admission to an exclusive club such as the EU) and sticks (such as arms embargoes), may push governments to engage in reform initiatives, or weaken those that do not. Equally, the absence of such external factors can encourage internal resistance to reform.

Ultimately, however, military reform cannot be effective if it is not driven from within. This requires a change of minds and perceptions within society, as well as within the military. The criticism of the TNI that built up within Indonesian society and within the military itself during the New Order years, which eventually led to the breaking of traditional taboos against challenging the military, represents a more bottom-up process than do Turkey's EU-imposed democratic reforms. This explains the greater levels of public enthusiasm for Indonesia's military reforms during the *reformasi* than were in evidence for Turkey's military reforms.

Economic development is necessary, but not sufficient
When a significant proportion of a country's population is preoccupied with the task of daily survival, it is unreasonable to expect genuine

popular passion about the ideals of democratisation and military reform. Proponents of 'genuine' democratisation tend to be university students, scholars, intellectuals and small business owners – i.e., members of the middle class. Economic development and the eradication of extreme poverty, and the expansion of the middle class, which contributes to reducing the income discrepancies that plague most developing countries, are thus instrumental to the consolidation of democratic reforms.

That being said, economic development alone does not ensure the establishment of a substantive democracy. As evidenced by both our case studies, reforms that are primarily driven by the motivation of achieving economic development risk being superficial, as the internal drive for reform usually wanes when short-term economic conditions change. Substantive democratisation necessitates the evolution of a democratic political culture in society at large. Admittedly, this is one of the most paradoxical aspects of democratisation, as which factor precedes the other – whether economic development or the evolution of a democratic political culture comes first – is not always easy to determine.

Greater professionalism leads to more efficiency and less corruption in the military, but not necessarily to more democratisation

Regarding the classic debate in civil–military relations concerning whether greater professionalism within the military makes for a more or less politically involved military, the evidence of the militaries discussed above would seem to suggest that, while greater professionalism can make a military more disciplined, efficient and resistant to corruption (as per Samuel Huntington), it does not, as Morris Janowitz warned, discourage the military from interfering in civilian affairs. Better trained and better paid, the Turkish TSK has a brighter track record than the Indonesian TNI in terms of discipline, efficiency and 'ethical' behaviour. At the same time, this track record only seems to reinforce the Turkish military's – and the Turkish people's – perception of the TSK as the rightful owner of the state, and increase a generalised lack of confidence in elected civilian authorities, giving all the political leeway to the military that this would indicate.

Evidence supports the 'third wave' theory of democratisation, and points to the coming of a reverse wave

The recent experience of the two countries discussed supports Huntington's theory of a 'wave' of democratisation sweeping through the world, carried to a large extent by globalisation. It has been our argument that the Turkish and Indonesian reform initiatives of the late 1990s and the early 2000s were heavily influenced by the profound socio-political and economic impacts of globalisation.

At the same time, since 11 September 2001, and Washington's 'war on terror', democratisation has been in retreat, overtaken by a new drive to securitisation and the re-emerging prominence of security-sector actors. Furthermore, as we have seen, widespread opposition, especially in Muslim societies, to the US occupation of Iraq and Afghanistan in the name of democracy has resulted in a general alienation from and scepticism about the concept of democratisation in many places. The fact that these processes can be seen in both Turkey and Indonesia, two allies of the US, located in entirely different geopolitical environments, supports the argument that a 'reverse wave' may be on its way.

Practical realism, rather than top-down idealism, is needed in the evaluation and promotion of military reform and democratisation processes

While the goal of democratisation and military reform may be evident – to make it exclusively the task of the people, via democratically elected civilian authorities, to determine their own national fate – there is clearly no fixed formula for achieving this ideal result. Every country presents unique qualities that affect whether democratisation is promoted or resisted. Realistic analyses of a country's reform process will take into account these specific cultural, historical, socio-political, economic and international determinants, and be wary of judging a country's progress against idealised normative expectations.

Both the countries examined here display a deep-rooted cultural inclination to place the military in a more revered social position than it enjoys in most Western democracies. Expecting and demanding that the military will abandon this position, and that societies will change their cultural characteristics overnight, may not only prove to be futile, but could also help a situation to deteriorate: militaries, states and societies can all be notoriously resistant, even hostile, to democratic change when they perceive it as a fundamental challenge to their cultural *raison d'être*. Besides, the impulse to impose change from the outside – however noble and altruistic the cause – has much in common with the arrogant rationale of those militaries that believe only they know what is best for 'the people'.

Instead, when evaluating a country's reform process, or when taking part in it, setting practical, limited goals is the most productive approach. It should be made abundantly clear to all actors in comparable reform situations that democratisation is not 'against' the military, but will ultimately benefit it. In the case of Turkey, the primary focus of reform initiatives, and the measure of their success, should be the dismantling of the authoritarian constitutional system put in place after the 1980 coup, rather than pressuring the military to abandon the guardianship role conferred on it by Kemalism, which it will not abandon. A new and a more democratic constitution will make it harder for the military to justify its interventions in civilian affairs, as these actions will no longer have a constitutional basis. In the case of Indonesia, the first goal of reformers should be to persuade the TNI that overcoming the authoritarian legacy of the New Order regime will in fact help to sustain *Pancasila*, and that it will consolidate, rather than damage, the TNI's authority, as it is the only way the discredited institution's reputation may be repaired.

The success of any reforming project will ultimately be determined by how much it empowers civil society – through economic development, social justice, education, respect for civil liberties, the free debate of controversial issues – rather than by the effects of imposed changes that the population does not fully support or demand. This is the criterion by which the long-term results of a country's military reforms and democratisation should be judged.

NOTES

Chapter One

1. Timothy Edmunds, 'Security Sector Reform: Concepts and Implementation', report presented to the Geneva Centre for Democratic Control of Armed Forces workshop 'Security Sector Reform: Conceptual Framework and Practical Implications', Geneva, 20–22 November 2001, p. 5.

2. Jean Grugel, *Democratisation: A Critical Introduction* (Basingstoke: Palgrave Macmillan, 2002), p. 2.

3. *Ibid.*, p. 17.

4. Laurence Whitehead, 'International Aspects of Democratization', in Guillermo O'Donnell, Philippe C. Schmitter and Laurence Whitehead (eds), *Transition from Authoritarian Rule: Comparative Perspectives* (Baltimore, MD: Johns Hopkins University Press, 1991), p. 8.

5. *Ibid.*, p. 9.

6. Juan J. Linz and Alfred Stepan, *Problems of Democratic Transition and Consolidation* (Baltimore, MD: Johns Hopkins University Press, 1996), p. 3.

7. Samuel Huntington, *The Third Wave: Democratisation in the Late Twentieth Century* (Norman, OK: University of Oklahoma Press, 1991), p. 10.

8. David Held, *Democracy and the Global Order: From the Modern State to Cosmopolitan Governance* (Cambridge: Polity Press, 1996), p. 15.

9. Mary Kaldor and Ivan Vejvoda, 'Democratisation in Central and East European Countries', *International Affairs*, vol. 73, no. 1, January 1997, p. 62.

10. Grugel, *Democratisation: A Critical Introduction*, p. 5.

11. Hans Born, Philipp H. Fluri and Simon Lunn (eds), *Oversight and Guidance: The Relevance of Parliamentary Oversight for the Security Sector and its Reform* (Brussels/Geneva: DCAF, 2003), p. 13.

12. Huntington, *The Soldier and the State: The Theory and Politics of Civil–Military Relations* (Cambridge, MA: Belknap Press, 1957), p. 84.

13. Morris Janowitz, *The Professional Soldier: A Social and Political Portrait* (New York: The Free Press, 1960).

14. Asha Gupta (ed.), *Military Rule and Democratisation: Changing Perspectives* (New Delhi: Deep and Deep Publications, 2003), p. 6.

15. Born, Fluri and Lunn, *Oversight and Guidance: The Relevance of Parliamentary Oversight for the Security Sector and its Reform*, p. 14.

16. Edmunds, 'Security Sector Reform: Concepts and Implementation', pp. 1–2.

17. Grugel, *Democratisation: A Critical Introduction*, p. 91.

18. Huntington, *The Third Wave: Democratisation in the Late Twentieth Century*, p. 5.

19 Linz and Stepan, *Problems of Democratic Transition and Consolidation*, p. 66.

20 Theodor H. Winkler, 'Managing Change: The Reform and Democratic Control of the Security Sector and International Order', Geneva Centre for Democratic Control of Armed Forces, Occasional Paper no. 1, Geneva, October 2002, p. 8.

21 Grugel, *Democratisation: A Critical Introduction*, p. 1.

22 *Ibid.*, p. 67.

23 Emily Goldman, 'Cultural Foundations of Military Diffusion', *Review of International Studies*, no. 32, 2006, p. 70.

24 Yezid Sayigh, 'Military and State Relationship in the Middle East', lecture in research seminar, Centre of International Studies, University of Cambridge, 11 October 2007.

25 A regional approach, as Morris Janowitz has argued, has proven particularly productive for the analysis of civil–military relations. Janowitz, *Civil Military Relations: Regional Perspectives* (London: Sage, 1981), p. 9.

26 Huntington, *The Third Wave: Democratisation in the Late Twentieth Century*, pp. 15–26.

27 *Ibid.*, pp. 45–6.

28 Francis Fukuyama, *The End of History and the Last Man* (London: Hamish Hamilton, 1992).

29 Huntington, *The Third Wave: Democratisation in the Late Twentieth Century*, pp. 29–30.

30 'Syria: Has He Got Away With It?', *The Economist*, 7 April 2007.

31 Hans H. Gerth and C. Wright Mills (eds), *From Max Weber: Essays in Sociology* (Oxford: Oxford University Press, 1958), pp. 77–8.

32 Born, Fluri and Lunn, *Oversight and Guidance: The Relevance of Parliamentary Oversight for the Security Sector and its Reform*, p. 30.

33 Edmunds, 'Security Sector Reform: Concepts and Implementation', p. 6.

34 Huntington, *The Third Wave: Democratisation in the Late Twentieth Century*, p. 232.

35 *Ibid.*, pp. 233–43.

36 *Ibid.*, pp. 6–7.

37 Edmunds, 'Security Sector Reform: Concepts and Implementation', pp. 9–10.

Chapter Two

1 Suna Kili, 'Role of the Military in Turkish Society: An Assessment from the Perspective of History, Sociology and Politics', in Gupta, *Military Rule and Democratisation: Changing Perspectives*, p. 145.

2 Ümit Cizre, 'The Anatomy of the Turkish Military's Political Autonomy', *Comparative Politics*, vol. 29, no. 2, January 1997, p. 154.

3 Mevlüt Bozdemir, *Türk Ordusunun Tarihsel Kaynakları* (Ankara: AÜ Siyasal Bilgiler Fakültesi Yayınları, 1982), p. 157.

4 Tanel Demirel, 'Türk Silahlı Kuvvetleri'nin Toplumsal Meşruiyeti Üzerine', in Ahmet İnsel and Ali Bayramoğlu (eds), *Bir Zümre, Bir Parti: Türkiye'de Ordu* (Istanbul: Birikim Yayınları, 2004), p. 348.

5 İnsel 'Bir Toplumsal Sınıf Olarak Türk Silahlı Kuvvetleri', in İnsel and Bayramoğlu, *Bir Zümre, Bir Parti: Türkiye'de Ordu*, p. 43.

6 Eduard Soler i Lecha, Débora Miralles i Solé, Ümit Cizre and Volkan Aytar, 'Drawing Lessons from Turkey's and Spain's Security Sector Reforms for the Mediterranean', EuroMesCo Research Project, Istanbul, October 2006.

7 Linda Michaud-Emin, 'The Restructuring of the Military High Command in the Seventh Harmonization Package and its Ramifications for Civil–Military Relations in Turkey', *Turkish Studies*, vol. 8, no. 1, March 2007, p. 33.

8 İnsel and Bayramoğlu, *Bir Zümre, Bir Parti: Türkiye'de Ordu*, p. 50.

9   İlhan Uzgel, 'Ordu Dış Politikanın Neresinde?', in İnsel and Bayramoğlu, *Bir Zümre, Bir Parti: Türkiye'de Ordu*, p. 325.

10   For example, in a nationwide survey conducted by the governing AKP in January 2005, 84% of respondents said that they trusted the TSK more than any other institution in the country. The same survey found that political parties and the media were among the least trusted institutions. *Hürriyet*, 18 January 2005.

11   Demirel, 'Türk Silahlı Kuvvetleri'nin Toplumsal Meşruiyeti Üzerine', p. 353.

12   Ahmet Taner Kışlalı, 'Türk Ordusunun Toplumsal Kökeni Üzerine bir Araştırma', *Siyasal Bilgiler Fakültesi Dergisi*, Ankara University, no. 3–4, September–December 1974, p. 90.

13   Metin Heper, 'The Military–Civilian Relations in Post-1997 Turkey', in George Cristian Maior and Larry Watts (eds), *Globalization of Civil–Military Relations: Democratization, Reform and Security* (Bucharest: Enciclopedica Publishing House, 2002), p. 58.

14   Bulent Aliriza, 'Turks Have an Unavoidable War to Fight Against Corruption', *International Herald Tribune*, 16 March 2001.

15   Ayşe Gül Altınay, 'Eğitimin Militarizasyonu: Zorunlu Milli Eğitim Dersi', in İnsel and Bayramoğlu, *Bir Zümre, Bir Parti: Türkiye'de Ordu*, pp. 179–201.

16   Ali Bayramoğlu, Ahmet İnsel and Ömer Laçiner, 'Giriş', in İnsel and Bayramoğlu, *Bir Zümre, Bir Parti: Türkiye'de Ordu*, p. 8.

17   'Turkey's Military Leader Vows to Restore Democracy', *New York Times*, 6 January 1981.

18   Heper, 'The Military–Civilian Relations in Post-1997 Turkey', p. 52.

19   Ethem Ruhi Fığlalı, 'Islam ve Laiklik', *Atatürk Araştırma Merkezi Dergisi*, no. 33, vol. 11, November 1995, p. 653.

20   Heper, 'The Military–Civilian Relations in Post-1997 Turkey', p. 60.

21   Linz and Stepan, 'Towards Consolidated Democracies', *Journal of Democracy*, vol. 7, no. 2, 1996, p. 16.

22   Mark Tessler and Ebru Altınoğlu, 'Political Culture in Turkey: Connections among Attitudes Towards Democracy, the Military and Islam', *Democratization*, vol. 11, no. 1, February 2004, p. 44.

23   *Ibid.*, p. 45.

24   Gül Altınay, 'Eğitimin Militarizasyonu: Zorunlu Milli Eğitim Dersi', p. 186.

25   The positive effects of the increase in Turkey's GDP per capita in recent years (from $7,000 in 2003 to $8,900 in 2006) have been limited by increasing income discrepancies. According to the UN Development Programme's 2006 Human Development Report, 18.7% of Turkey's population lives on less than $2 a day. See Bosphorus University economics department, 'Income Distribution' presentation, http://www.econ.boun. edu.tr/courses/spring2007/EC470_01/ readings/11_Income%20Distribution. ppt

26   Demirel, 'Türk Silahlı Kuvvetleri'nin Toplumsal Meşruiyeti Üzerine', p. 353.

27   Ece Temelkuran, 'Şeriat Mı Gelsin, Ordu Mu?', *Milliyet*, 6 April 2007.

28   Ali Karaosmanoğlu and Seyfi Tashan, *The Europeanization of Turkey's Security Policy: Prospects and Pitfalls* (Ankara: Foreign Policy Institute, 2004), p. 20.

29   Gencer Özcan, 'National Security Council', in Cizre, *Almanac Turkey 2005: Security Sector and Democratic Oversight* (Istanbul: TESEV, 2006), pp. 43–4.

30   İnsel and Bayramoğlu, *Bir Zümre, Bir Parti: Türkiye'de Ordu*, p. 49.

31   Defence spending routinely amounts to more than 3% of Turkey's GDP (it varied between 10.9% and 8.2% of the budget between 1997 and 2004). The TSK also has various extra-budgetary resources, including the Armed Forces Pensions Fund and the Defence Industry Support Fund. Pınar Akkoyunlu, *Eğitim ve Ekonomi* (Istanbul: Filiz, 2005), p. 183.

32   Karaosmanoğlu and Tashan, *The Europeanization of Turkey's Security Policy: Prospects and Pitfalls*, p. 11.

33   Cizre, *Almanac Turkey 2005: Security Sector and Democratic Oversight*, p. 13.

34 Karaosmanoğlu and Tashan, *The Europeanization of Turkey's Security Policy: Prospects and Pitfalls*, p. 12.

35 For the Turkish contribution to the International Security Assistance Force, see http://www.nato.int/isaf/structure/regional_command/index.html.

36 For Turkey's role in Peace Support Operations, see http://www.tsk.mil.tr/eng/uluslararasi/barisdestekkatki.htm, and in the Partnership for Peace, http://www.bioem.tsk.mil.tr/.

37 Feroz Ahmad, *Turkey: The Quest for Identity* (Oxford: Oneworld, 2005), pp. 177, 184.

38 Lale Sarıibrahimoğlu, 'Turkish Armed Forces', in Cizre, *Almanac Turkey 2005: Security Sector and Democratic Oversight*, p. 63.

39 'Kılınç Paşa NATO'dan Çıkma Çağrısı Yaptı', *Sabah*, 29 May 2007; 'Military Manoeuvres', *The Economist*, 7 June 2007.

40 Karaosmanoğlu and Tashan, *The Europeanization of Turkey's Security Policy: Prospects and Pitfalls*, p. 19.

41 Sebnem Arsu, 'Turkey Angry Over House Armenian Genocide Vote', *New York Times*, 12 October 2007.

42 Stephen Larrabee and Ian Lesser, *Turkish Foreign Policy in an Age of Uncertainty* (Washington DC: RAND Publications, 2003), p. 52.

43 İnsel and Bayramoğlu, *Bir Zümre, Bir Parti: Türkiye'de Ordu*, p. 108.

44 Academic Survey Centre, *Türkiye ve Avrupa Birliği İlişkileri Araştırması* (Istanbul: AKART, May 2002). 29% of respondents said EU membership was needed to democratise Turkey and modernise the state, while 30% supported Turkey's EU bid because they thought membership would improve Turkey's economic welfare.

45 Over 80% of the respondents to the 2002 AKART poll were in favour of EU membership. However, more than half of these also stated that Turkey should join without making any strategic concessions. What Brussels sees as democratisation, many people in Turkey interpret as strategic concessions.

46 Mustafa Acar, 'Avrupa Birliği'ne Tepkiler: Türkiye'nin Daha İyi bir Alternatifi var Mı?', *CÜ İktisadi ve İdari Bilimler Dergisi*, vol. 2, no. 2, 2002, p. 75.

47 The rate of inflation in 2002 was 29.7%; in 2005, it had fallen to 7.7%, the lowest rate in 37 years. In the same period, GNP and per capita income almost doubled: in 2002, GNP stood at $180bn, and GNP per capita at $2,598; in 2005, GNP was $360bn, and GNP per capita $5,008. *Annual Report 2006*, Turkish Ministry of Finance, http://www.sgb.gov.tr/eKtphane/2006yer.pdf.

48 'Halk AB'ye Güvenmiyor', *Milliyet*, 24 October 2006.

49 European Commission, *Turkey 2005 Progress Report*, Brussels, 9 November 2005, p. 14, http://ec.europa.eu/enlargement/archives/pdf/key_documents/2005/package/sec_1426_final_progress_report_tr_en.pdf.

50 'EU Warns Turkish Army Over Vote', BBC News, 28 April 2007, http://news.bbc.co.uk/1/hi/world/europe/6602661.stm.

51 Huntington, *The Third Wave: Democratisation in the Late Twentieth Century*, p. 283.

52 *Ibid.*

53 Heper, 'The Military–Civilian Relations in Post-1997 Turkey', p. 54.

54 Ivar Ekman, 'Top Swedish Official Backs Turkey for EU', *International Herald Tribune*, 11 December 2006.

55 Cizre, *Almanac Turkey 2005: Security Sector and Democratic Oversight*, p. 4.

56 İnsel and Bayramoğlu, *Bir Zümre, Bir Parti: Türkiye'de Ordu*, p. 74.

57 For details, see the TSK official website, http://www.tsk.mil.tr/.

58 Larrabee and Lesser, *Turkish Foreign Policy in an Age of Uncertainty*, p. 168.

59 This was later acknowledged by the CIA's Ankara chief Paul Henze, who, on the night of the coup, cabled Washington, saying 'our boys have done it'. Mehmet Ali Birand, *The Generals' Coup in Turkey: An Inside Story of 12 September 1980* (New York: Elsevier, 1987).

60 Gencer Özcan and Şule Kut, *En Uzun Onyıl: Türkiye'nin Ulusal Güvenlik ve*

Dış Politika Gündeminde Doksanlı Yıllar (Istanbul: Büke Yayınları, 2000), p. 17.

61   An indication of current opinion of the US in Turkey is given by a 2007 Pew Global Attitudes survey, which found only 9% of respondents in Turkey looking favourably upon the US, down dramatically from 52% before the invasion of Afghanistan in 2001. The survey also found intensified distrust of America in many other parts of the world. Meg Bortin, 'Global Poll Shows Wide Distrust of United States', *International Herald Tribune*, 27 June 2007, http://www.iht.com/articles/2007/06/27/news/pew.php?page=1.

62   Uzgel, 'Ordu Dış Politikanın Neresinde?', p. 74.

63   Cüneyt Ülsever, 'George W. Bush: Ortadoğu Için Tarihi Bir Konuşma', *Hürriyet*, 1 July 2004. Rather than aiding Turkey with the accession process, however, these expressions of support have arguably led to the shaping of a perception in the core member states of the EU, especially France and Germany, of Turkey as an American Trojan horse through which Washington might extend its influence inside the EU. See Charles Grant, 'Turkey Offers EU More Punch', *European Voice*, 1–7 September 2005.

64   Nikolaos Raptopoulos, 'Rediscovering its Arab Neighbours? The AKP Imprint on Turkish Foreign Policy in the Middle East', *Les Cahiers du RMES*, no. 1, July 2004, p. 12.

65   'US Discourages Turkish Military Action in Northern Iraq', *Turkish Daily News*, 29 June 2007.

66   'Demokrasi Kesintiye Uğramasın', *Hürriyet*, 29 April 2007.

67   Ahmad, *Turkey: The Quest for Identity*, p. 132.

68   Cizre, 'The Anatomy of the Turkish Military's Political Autonomy', p. 153.

69   *Ibid.*, p. 157.

70   Tessler and Altınoğlu, 'Political Culture in Turkey: Connections among Attitudes Towards Democracy, the Military and Islam', p. 24.

71   Larrabee and Lesser, *Turkish Foreign Policy in an Age of Uncertainty*, p. 63.

72   The 1997 National Security Policy Document identified political Islam as the main threat to the secular republic. A few months later, the Islamist government was toppled in a coup.

73   Bayramoğlu, 'Asker ve Siyaset', in İnsel and Bayramoğlu, *Bir Zümre, Bir Parti: Türkiye'de Ordu*, p. 92.

74   Article 2a of Law no. 2945, on the National Security Council, defines national security as 'the protection of the constitutional order of the state, its nation and integrity, all of its interests in the international sphere including political, social, cultural and economic interests, as well as the protection of its constitutional law against all internal and external threats'. European Commission, *Turkey 2005 Progress Report*, p. 14. For the full text of the law in Turkish, see http://fef.comu.edu.tr/sivilsavunma/kanunlar/ka2945mgk.pdf.

75   Soler i Lecha, Miralles i Solé, Cizre and Aytar, 'Drawing Lessons from Turkey's and Spain's Security Sector Reforms for the Mediterranean', p. 11.

76   Ilkay Sunar and Sabri Sayari, 'Democracy in Turkey: Problems and Prospects', in Guillermo O'Donnell, Philippe C. Schmitter and Laurence Whitehead (eds), *Transition from Authoritarian Rule: Southern Europe* (Baltimore, MD: Johns Hopkins University Press, 1986), p. 184.

77   *Ibid.*

78   Huntington, *The Third Wave: Democratisation in the Late Twentieth Century*, pp. 238–9.

79   Ahmad, *Turkey: The Quest for Identity*, p. 151.

80   Larrabee and Lesser, *Turkish Foreign Policy in an Age of Uncertainty*, p. 52.

81   Hakan M. Yavuz (ed.), *The Emergence of New Turkey: Democracy and the AK Parti* (Salt Lake City, UT: Utah University Press, 2006), pp. 1–23.

82   Soler i Lecha, Miralles i Solé, Cizre and Aytar, 'Drawing Lessons from Turkey's and Spain's Security Sector Reforms for the Mediterranean', p. 13.

83   For details of the National Security Council reforms, see Özcan, 'National

Security Council', in Cizre, *Almanac Turkey 2005: Security Sector and Democratic Oversight*, pp. 36–50.

84 Sarıibrahimoğlu, 'Turkish Armed Forces', in Cizre, *Almanac Turkey 2005: Security Sector and Democratic Oversight*, p. 60. The word 'reported' is used here because the document is not made public. The only way the public is informed about its content is through leaks to the press.

85 *Ibid.*, p. 58.

86 Larrabee and Lesser, *Turkish Foreign Policy in an Age of Uncertainty*, p. 63.

87 İnsel, 'Bir Toplumsal Sınıf Olarak Türk Silahlı Kuvvetleri', p. 46.

88 Avni Özgürel, 'En Milliyetçi: Erdoğan ve AKP!..', *Radikal*, 14 March 2007.

89 'Turkish Military Chief Flexes Some Political Muscle', *Financial Times*, 27 February 2007; 'Paşa'dan Sürpriz Basın Toplantısı', *Sabah*, 12 April 2007.

90 'Askerin Medya Notları', *Radikal*, 8 March 2007.

91 'İçinden iki Darbe Girişimi Geçen Günlük', *Radikal*, 29 March 2007.

92 'Nokta Muhabiri İçin Hapis Istemi', CNN Türk, 22 April 2007, http://www.cnnturk.com/TURKIYE/haber_detay.asp?PID=318&HID=1&haberID=351145.

93 Perihan Mağden, 'Basın Özgürlüğünde son Nokta', *Radikal*, 17 April 2007.

94 Karabekir Akkoyunlu, 'Caught in the Middle of Turkey's Fundamentalisms', *Turkish Daily News*, 16 June 2007.

95 'Turks Elect Ex-Islamist President', BBC News, 28 August 2007, http://news.bbc.co.uk/1/hi/world/europe/6966216.stm.

96 Ertuğrul Özkök, '2nci Cumhuriyetin 1nci Cumhurbaşkanı', *Hürriyet*, 29 August 2007.

97 See Bekir Coşkun, 'Düğün Gecesi', *Hürriyet*, 1 September 2007.

98 'Democracy in Turkey', *New York Times*, 1 September 2007.

99 In an encouraging sign, a group of academics appointed by the government for the purpose of writing a new constitution had begun work on a draft of a new and 'more civilian' one as early as September 2007. How much the new constitution will help improve Turkey's democracy remains, nevertheless, to be seen. *Turkish Daily News*, 4 September 2007.

100 Hasan Cemal, 'Bu da Milletin Muhtırası!', *Milliyet*, 23 July 2007.

101 The murders of a Catholic priest in Trabzon, three Protestant publishers of the Bible in Malatya, and Hrant Dink, a vocal Turkish–Armenian journalist, in Istanbul during the first half of 2007, are cases in point.

102 'Turkey and Transatlantic Trends: Between Xenophobia and Globalization', *Transatlantic Trends 2007*, http://www.transatlantictrends.org.

103 'Turkish MPs Back Attacks in Iraq', BBC News, 18 October 2007, http://news.bbc.co.uk/1/hi/world/europe/7049348.stm.

104 Amberin Zaman, 'Turkey Elects Islamist President Abdullah Gul', *Daily Telegraph*, 31 August 2007, http://www.telegraph.co.uk/news/main.jhtml?xml=/news/2007/08/29/wturkey129.xml.

Chapter Three

1 *CIA World Factbook*, https://www.cia.gov/library/publications/the-world-factbook/geos/id.html.

2 Dewi Fortuna Anwar, 'Negotiating and Consolidating Democratic Civilian Control of the Indonesian Military', East–West Center Occasional Papers, Politics and Security Series, no. 4, February 2001, p. 9.

3 Robert Lowry, *The Armed Forces of Indonesia* (Sydney: Allen and Unwin, 1996), p. 192.

4 Damien Kingsbury, *The Politics of Indonesia* (Oxford: Oxford University Press, 2002), p. 275.

5   This last clause has become one of the main ideological tools employed by the military to justify its socio-political role and excessive political influence.

6   Pratikno, 'Exercising Freedom: Local Autonomy and Democracy in Indonesia, 1999–2001', in Maribeth Erb, Priyambudi Sulistiyanto and Carole Faucher (eds), *Regionalism in Post-Suharto Indonesia* (London: RoutledgeCurzon, 2005), p. 23.

7   Anwar, 'Negotiating and Consolidating Democratic Civilian Control of the Indonesian Military', p. 8.

8   BICC Country Briefing, 'Security Sector Reform in Indonesia', Bonn International Centre for Conversion, 2005, p. 2.

9   Lowry, *The Armed Forces of Indonesia*, p. 192.

10 Pratikno, 'Exercising Freedom: Local Autonomy and Democracy in Indonesia, 1999–2001', p. 23.

11 Lowry, *The Armed Forces of Indonesia*, p. 178.

12 *Ibid.*, p. 148.

13 *Ibid.*, p. 196.

14 In the late 1980s, Suharto helped to form and supported the Indonesian Association of Islamic Intellectuals; he also cultivated a 'green', or Islamist, faction within the military against the 'red and white' – nationalist secular – hardline officers. For the Indonesian Association of Islamic Intellectuals, see Angel Rabasa and John Haseman, *The Military and Democracy in Indonesia: Challenges, Politics, and Power* (Washington DC: RAND Monograph Report, 2002), p. 37. For the 'green faction', see Kingsbury, *The Politics of Indonesia*, p. 277.

15 Lowry, *The Armed Forces of Indonesia*, p. 201. Emphasis added.

16 The 'New Paradigm' was subsequently published as a book entitled 'The TNI in the Twenty-First Century'. DEPHANKAM, *TNI Abad XXI: Redefinisi, Reposisi, dan Reaktualisisi Peran TNI dalam Kehidupan Bangsa* (Jakarta: Jasa Buma, 1999).

17 Jun Honna, *Military Politics and Democratization in Indonesia* (New York: RoutledgeCurzon, 2006), p. 165.

18 Anwar, 'Negotiating and Consolidating Democratic Civilian Control of the Indonesian Military', p. 1.

19 Lex Reiffel, 'Indonesia's Quiet Revolution', *Foreign Affairs*, vol. 83, no. 5, 2004, p. 104.

20 Kingsbury, *The Politics of Indonesia*, p. 290.

21 Anwar, 'Negotiating and Consolidating Democratic Civilian Control of the Indonesian Military', p. 26.

22 Aceh is rich in oil and gas. Irian Jaya is the site of the world's largest gold and copper mining operations; it also has extensive oil and gas reserves. Rabasa and Haseman, *The Military and Democracy in Indonesia: Challenges, Politics, and Power*, p. 106.

23 Lowry, *The Armed Forces of Indonesia*, p. 178.

24 For more on *oknum-oknum*, see *ibid.*, p. 179.

25 The military established a 'military operation zone' in Aceh province in 1990 in response to the increasing number of uprisings there in the late 1980s. Dewi Fortuna Anwar, 'Negotiating and Consolidating Democratic Civilian Control of the Indonesian Military', p. 22.

26 Rabasa and Haseman, *The Military and Democracy in Indonesia: Challenges, Politics, and Power,* p. 69. The military's economic interests will be elaborated in the 'state' section.

27 BICC Country Briefing, 'Security Sector Reform in Indonesia', p. 3.

28 Lowry, *The Armed Forces of Indonesia*, p. 145.

29 Harold Crouch, *The Army and Politics in Indonesia* (Ithaca, NY: Cornell University Press, 1988), p. 285.

30 Reiffel, 'Indonesia's Quiet Revolution', p. 105.

31 Sukardi Rinakit, *The Indonesian Military After the New Order* (Singapore: NIAS Press, 2005), p. 61.

32 Adam Schwarz, *A Nation in Waiting: Indonesia's Search for Stability* (Boulder, CO: Westview Press, 2000), p. 399.

33 For more on *keterbukaan*, see Honna, *Military Politics and Democratization in Indonesia*, pp. 8–14.

34 *Ibid.*, pp. 109–10. Harold Crouch characterises the main divide as being between 'financial' and 'professional'

generals; the former wishing to preserve the status quo, the latter opting for reform. Crouch, *The Army and Politics in Indonesia*, pp. 308–9.

[35] Rabasa and Haseman, *The Military and Democracy in Indonesia: Challenges, Politics, and Power*, p. 35.

[36] Schwarz, *A Nation in Waiting: Indonesia's Search for Stability*, p. 42.

[37] As translated in Honna, *Military Politics and Democratization in Indonesia*, p. 54.

[38] Pratikno, 'Exercising Freedom: Local Autonomy and Democracy in Indonesia, 1999–2001', p. 23.

[39] Rabasa and Haseman, *The Military and Democracy in Indonesia: Challenges, Politics, and Power*, p. 30.

[40] Terence Lee, 'Indonesian Military Taking the Slow Road to Reform', *Straits Times*, 18 December 2004.

[41] 'When Generals go on Manoeuvres', *Jakarta Post*, 26 January 2007.

[42] Rinakit, *The Indonesian Military After the New Order*, p. 240.

[43] Ikrar Nusa Bhakti, 'The Transition to Democracy in Indonesia: Some Outstanding Problems', in Jim Rolfe (ed.), *The Asia Pacific: A Region in Transition* (Honolulu, HI: Asia-Pacific Center for Security Studies, 2004), p. 195.

[44] Major post-1998 conflicts and movements: conflict between the Dayaks and the Madurese in Kalimantan, and between Christians and Muslims in the Moluccas; the resurgence of the Free Aceh Movement in Aceh; and the insurgencies in East Timor and Irian Jaya.

[45] Bhakti, 'The Transition to Democracy in Indonesia: Some Outstanding Problems', p. 195. While Indonesia's GDP per capita grew significantly between 1998 ($1,066) and 2006 ($3,900) (see *CIA World Factbook*), extreme poverty continues to plague a very significant portion of the population, and the income discrepancy between rich and poor has widened. In 2006, more than 39m people (18% of the population) lived beneath the official poverty line of $16.80 per month, notwithstanding the overall increase in national wealth. 'Poverty

in Indonesia: Always with Them', *The Economist*, 14 September 2006.

[46] Anwar, 'Negotiating and Consolidating Democratic Civilian Control of the Indonesian Military', p. 1.

[47] *Ibid.*

[48] *Ibid.*, pp. 32–3.

[49] Vedi Hadiz, 'Reorganizing Political Power in Indonesia', in Erb, Sulistiyanto and Faucher, *Regionalism in Post-Suharto Indonesia*, p. 45.

[50] Crouch, *The Army and Politics in Indonesia*, p. 273.

[51] Anwar, 'Indonesian Domestic Priorities Define National Security', in Muthiah Alagappa (ed.), *Asian Security Practice, Material and Ideational Influences* (Stanford, CA: Stanford University Press, 1998) p. 475.

[52] Anwar, 'Indonesia's Role in ASEAN', paper presented to seminar on 'The ASEAN Experience and its Relevance for SAARC', Jawaharlal Nehru University, New Delhi, 1995, p. 3.

[53] Schwarz, *A Nation in Waiting: Indonesia's Search for Stability*, pp. 19–23. These purges also saw the expulsion of leftist officers from the military. *Ibid.*, p. 29.

[54] Rinakit, *The Indonesian Military After the New Order*, p. 103.

[55] *Ibid.*

[56] *Ibid.*, p. 104.

[57] *Ibid.*

[58] Anwar, 'The Impact of the Asian Financial Crisis on Democratisation and Political Transition in Asia', paper presented to the conference on 'Asia Pacific Security in a Time of Economic Recovery', Asia-Pacific Center for Security Studies, Honolulu, Hawaii, August 1999, p. 3.

[59] Indonesia's economy shrank by 13.5% in 1997. The rupiah plunged from an exchange rate of 2,000 rupiah to $1 to 18,000 to $1 in a few months. http://www.imf.org/external/np/exr/facts/asia.pdf.

[60] Anwar, 'The Impact of the Asian Financial Crisis on Democratisation and Political Transition in Asia', pp. 5–7.

[61] Joakim Ojendal, 'Back to the Future? Regionalism in South-East Asia Under

Unilateral Pressure', *International Affairs*, vol. 80, no. 3, 2004, p. 522.

[62] Anwar, 'Indonesia's Role in ASEAN', p. 4.

[63] Ojendal, 'Back to the Future? Regionalism in South-East Asia Under Unilateral Pressure', p. 524.

[64] Anwar, 'The Future of Asia in the 21st Century and the Role of ASEAN', paper presented to the international symposium on 'Interaction for Progress: Myanmar in ASEAN', Yangon, Myanmar, October 1998, p. 4.

[65] 'Malaysian Foreign Minister Says ASEAN is No "Talk Shop"', *Asian Political News*, 12 May 2005.

[66] 'Lessons from Indonesia: Predatory Power Possible Under Democracy', *Jakarta Post*, 20 December 2006.

[67] The EU did initially refuse to conduct free-trade negotiations with ASEAN because of concerns about undemocratic regimes in certain member states. But talks were launched in February 2007, sparking protest among pro-democracy groups; see http://www.eubusiness.com/Trade/1178276401.21/.

[68] Gareth Evans, 'Indonesia's Military Culture has to be Reformed', *International Herald Tribune*, 24 July 2001.

[69] *Ibid.*

[70] 'Indonesia Protests Australia's Papua Visa Decision', ABC News Online, 24 March 2006, http://www.abc.net.au/news/newsitems/200603/s1600102.htm.

[71] Rabasa and Haseman, *The Military and Democracy in Indonesia: Challenges, Politics, and Power*, p. 113.

[72] BICC Country Briefing, 'Security Sector Reform in Indonesia', p. 6.

[73] Rabasa and Haseman, *The Military and Democracy in Indonesia: Challenges, Politics, and Power,* p. 114.

[74] Ojendal, 'Back to the Future? Regionalism in South-East Asia Under Unilateral Pressure', p. 533.

[75] BICC Country Briefing, 'Security Sector Reform in Indonesia', p. 6.

[76] 'Wolfowitz Dukung Kerja Sama Militer dengan Indonesia', *Kompas* (Indonesian daily), 11 October 2002.

[77] Tim Johnston, 'US Eases Indonesia Arms Ban', BBC News, 26 May 2005, http://news.bbc.co.uk/1/hi/world/asia-pacific/4581733.stm.

[78] 'IMET Resumption Seen as Recognition of TNI Reform', *Jakarta Post*, 1 March 2005.

[79] Anak Agung Banyu Perwita, 'Terrorism, Democratization and Security Sector Reform in Indonesia', working paper presented to Indonesia Research Unit, Institute of Political Science, Justus-Liebig University, Giessen, Germany, 20 December 2005, p. 12.

[80] Bhakti, 'The Transition to Democracy in Indonesia: Some Outstanding Problems', p. 205.

[81] Anwar, 'Negotiating and Consolidating Democratic Civilian Control of the Indonesian Military', p. 13.

[82] For the most recent official record of the composition of the MPR, as well as an explanation of its functions, see People's Consultative Assembly – Republic of Indonesia (MPR-RI), *Annual Report 2003*, http://www.mpr.go.id/pdf/ketetapan/putusan%20MPRRI%202003.pdf.

[83] Anwar, 'Negotiating and Consolidating Democratic Civilian Control of the Indonesian Military', p. 13.

[84] *Ibid.* In 1997, under mounting popular pressure for liberal reform, the Suharto government reduced the military contingent in the DPR to 75.

[85] BICC Country Briefing, 'Security Sector Reform in Indonesia', p. 2.

[86] Dwight Y. King, *Half-Hearted Reform: Electoral Institutions and the Struggle for Democracy in Indonesia* (Westport, CT: Praeger, 2003), p. 225.

[87] BICC Country Briefing, 'Security Sector Reform in Indonesia', p. 3.

[88] Anwar, 'Negotiating and Consolidating Democratic Civilian Control of the Indonesian Military', p. 19.

[89] In 2001, for instance, it amounted to 0.8% of GDP, and less than 4% of the government's budget. Husain Pontoh Coen, 'Guns, Ammunition and the Stench of Blood: Unravelling Military Involvement in the Ambon Conflict',

Pantau, year 3, no. 33, January 2003, p. 141.

90 Rabasa and Haseman, *The Military and Democracy in Indonesia: Challenges, Politics, and Power,* p. 70.

91 An audit of the main army foundation found that in 2000 its companies returned a net loss of 8.21bn rupiah ($985,000). Human Rights Watch, 'Too High A Price: The Human Rights Cost of the Indonesian Military's Economic Activities', vol. 18, no.5 (C), June 2006, p. 15.

92 Lowry, *The Armed Forces of Indonesia,* p. 144.

93 'Daerah Konflik, Peluang Bisnis Militer-Pengusaha', *Kompas,* 13 September 2003.

94 Ichsan Malik, 'Military Business in the Maluku Conflict Area', in Roem Topatimasang and Don K. Marut (eds), *The Land of Soldiers: Exploring the Political Economy of Indonesian Military* (Yogyakarta: INSIST Press, 2007), p. 131.

95 Human Rights Watch, 'Too High A Price: The Human Rights Cost of the Indonesian Military's Economic Activities', p. 72.

96 Danang Widoyoko, Irfan Muktiono, Adnan Topan Husodo, Barly Harliem Noe and Agung Wijaya, 'Bisnis Militer Mencari Legitimasi', Indonesia Corruption Watch, Jakarta, 2002, pp. 39–40, http://www.antikorupsi.org/docs/bukubismil.pdf.

97 *Kompas,* 7 October 2001, quotation translated in George Junus Aditjondro, 'Ebony, Security Post Business, Arms Trade, and Protecting Big Capital: The Political Economy of Military Business in Eastern Sulawesi', in Topatimasang and Marut, *The Land of Soldiers: Exploring the Political Economy of Indonesian Military,* p. 153.

98 'Presidential Push Needed on TNI's Internal Reform', *Jakarta Post,* 27 February 2007.

99 Ridwan Max Sijabat, 'Government Urged to Put Military Out of Business', *Jakarta Post,* 17 February 2007.

100 Anwar, 'Negotiating and Consolidating Democratic Civilian Control of the Indonesian Military', p. 16.

101 *Ibid.,* p. 17.

102 *Ibid.,* p. 20.

103 Perwita, 'Terrorism, Democratization and Security Sector Reform in Indonesia', p. 10.

104 Anwar, 'Negotiating and Consolidating Democratic Civilian Control of the Indonesian Military', p. 16.

105 Perwita, 'Terrorism, Democratization and Security Sector Reform in Indonesia', p. 10.

106 Anwar, 'Negotiating and Consolidating Democratic Civilian Control of the Indonesian Military', p. 26.

107 M. Taufiqurrahman, 'Defence Minister Says TNI Not Out to Regain New Order Power', *Jakarta Post,* 2 March 2007.

108 Kingsbury, *The Politics of Indonesia,* pp. 240–41.

109 Robert Elson, *Suharto: A Political Biography* (Cambridge: Cambridge University Press, 2001), p. 295.

110 Rinakit, *The Indonesian Military After the New Order,* p. 114.

111 Honna, *Military Politics and Democratization in Indonesia,* p. 39.

112 King, *Half-Hearted Reform: Electoral Institutions and the Struggle for Democracy in Indonesia,* p. 224.

113 Honna, *Military Politics and Democratization in Indonesia,* p. 181.

114 *Ibid.,* p. 178.

115 *Ibid.,* p. 179.

116 'Indonesia Not to Try Wiranto', Tempo Interaktive, 21 June 2007.

117 'Army Accused Over Moluccas Conflict', BBC News, 17 July 2000, http://news.bbc.co.uk/1/hi/world/asia-pacific/836852.stm.

118 Honna, *Military Politics and Democratization in Indonesia,* p. 183.

119 Kingsbury, *The Politics of Indonesia,* p. 297.

120 BICC Country Briefing, 'Security Sector Reform in Indonesia', p. 11.

121 *Ibid.,* p. 2.

122 Harry Bhaskara, 'Is Our Democracy on the Right Track?', *Jakarta Post,* 16 October 2006.

123 Rinakit, *The Indonesian Military After the New Order,* p. 234.

124 Human Rights Watch, 'Too High A Price: The Human Rights Cost of the Indonesian Military's Economic Activities', p. 19.

125 Alvin Darlanika Soedarjo, 'TNI Unveils New Doctrine: No Politics', *Jakarta Post,* 25 January 2007.

126 Rabasa and Haseman, *The Military and Democracy in Indonesia: Challenges, Politics, and Power*, p. 51.

127 Honna, *Military Politics and Democratization in Indonesia*, p. 200.

128 'Eighteen Regencies Sharia Bound', Indonesia Matters, 4 May 2006, http://www.indonesiamatters.com/320/eighteen-regencies-sharia-bound/.

129 John Aglionby, 'Jakarta Struggles with Politics of Pornography as *Playboy* Comes to Town', *Guardian*, 30 January 2006; 'A Softer Islamic Hard Line Sweeps Indonesia', *International Herald Tribune*, 29 June 2007.

ADELPHI PAPERS

The Adelphi Papers monograph series is the Institute's flagship contribution to policy-relevant, original academic research.

Eight Adelphi Papers are published each year. They are designed to provide rigorous analysis of strategic and defence topics that will prove useful to politicians and diplomats, as well as academic researchers, foreign-affairs analysts, defence commentators and journalists.

From the very first paper, Alastair Buchan's *Evolution of NATO* (1961), through Kenneth Waltz's classic *The Spread of Nuclear Weapons: More May Be Better* (1981), to influential additions to the series such as Mats Berdal's *Disarmament and Demobilisation after Civil Wars* (1996) and Lawrence Freedman's *The Transformation of Strategic Affairs* (2006), Adelphi Papers have provided detailed, nuanced analysis of key security issues, serving to inform opinion, stimulate debate and challenge conventional thinking. The series includes both thematic studies and papers on specific national and regional security problems. Since 2003, Adelphi Paper topics have included *Strategic Implications of HIV/AIDS*, *Protecting Critical Infrastructures Against Cyber-Attack*, *The Future of Africa: A New Order in Sight*, *Iraq's Future: The Aftermath of Regime Change*, *Counter-terrorism: Containment and Beyond*, *Japan's Re-emergence as a "Normal" Military Power*, *Weapons of Mass Destruction and International Order*, *Nuclear Terrorism After 9/11* and *North Korean Reform*.

Longer than journal articles but shorter than books, Adelphi Papers permit the IISS both to remain responsive to emerging strategic issues and to contribute significantly to debate on strategic affairs and the development of policy. While the format of Adelphi Papers has evolved over the years, through their authoritative substance and persuasive arguments recent issues have maintained the tradition of the series.

⊤IISS ADELPHI PAPERS

RECENT **ADELPHI PAPERS** INCLUDE:

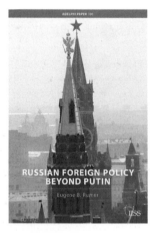
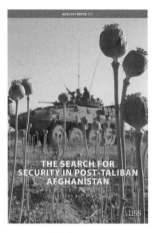